Abel Hernández-Muñoz

Wildlife of the Antilles

Abel Hernández-Muñoz

Wildlife of the Antilles

Cuba's insular Caribbean biodiversity hotspot for forestry engineering students

ScienciaScripts

Imprint

Any brand names and product names mentioned in this book are subject to trademark, brand or patent protection and are trademarks or registered trademarks of their respective holders. The use of brand names, product names, common names, trade names, product descriptions etc. even without a particular marking in this work is in no way to be construed to mean that such names may be regarded as unrestricted in respect of trademark and brand protection legislation and could thus be used by anyone.

Cover image: www.ingimage.com

This book is a translation from the original published under ISBN 978-620-2-16919-6.

Publisher:
Sciencia Scripts
is a trademark of
Dodo Books Indian Ocean Ltd. and OmniScriptum S.R.L publishing group

120 High Road, East Finchley, London, N2 9ED, United Kingdom
Str. Armeneasca 28/1, office 1, Chisinau MD-2012, Republic of Moldova, Europe
Printed at: see last page
ISBN: 978-620-7-55620-5

WILDLIFE OF THE ANTILLES

MSc. Abel Hernández Muñoz

Table of Contents

1. INTRODUCTION 3

2. BIOLOGICAL IMPORTANCE OF THE Biodiversity Hotspot of the West Indies 6

3. CONSERVATION OUTCOMES DEFINED FOR THE ANTILLES BIODIVERSITY HOTSPOT 30

4. THREATS TO WILDLIFE IN THE HOT SPOT 70

5. FAUNA OF TRINIDAD &TOBAGO, A SPECIAL CASE IN THE ANTILLES 106

1. INTRODUCTION

The islands of the Antilles are, in general, mountainous. In Cuba there are hundreds of caves excavated in the limestone rocks, the same happens in Hispaniola and Puerto Rico. The rivers, however, are scarce and short.

Volcanoes are abundant in the Lesser Antilles. Near the coast there are oceanic trenches where the sea reaches a great depth. But in many islands there are extensions of the insular platforms, which turns these coastal segments into beaches and wetlands.

The **climate is tropical.** They have warm temperatures and abundant rainfall during the rainy season (June to October). In addition, **hurricanes** form continuously in the area between August and November, which, if they approach the coast, cause considerable destruction.

The vegetation grows in great variety, with many exclusive and exuberant species on almost all the islands, providing great shelter and food for wildlife.

In general, the fauna varies from one island to another, although large mammals are scarce in all of them. Birds are very numerous: the **hummingbird** is abundant, of small size, and can move back and forth while tasting the nectar of the flower. Other common birds are **parrots** and **cuckoos,** as well as **flamingos** and many others. There are also many iguanas, bats, turtles...

Despite its small size, the Antilles support one of the highest rates of presence of threatened species of any biodiversity *hotspot* in the world. The Antilles is a biodiversity-rich archipelago comprising 30 countries and territories and extending over almost 4 million km^2 of sea (Figure 1). The Antilles is one of 36 biodiversity hotspots in the world. Biodiversity hotspots contain at least 1 500 plant species found nowhere else and have lost at least 70% of their original natural habitat (Mittermeier *et al.* 2004). The biogeography of the islands and the complex geology of the Caribbean have created unique habitats and a great diversity of species. In addition, these islands support one of the highest occurrence rates of threatened species of any hotspot in the world.

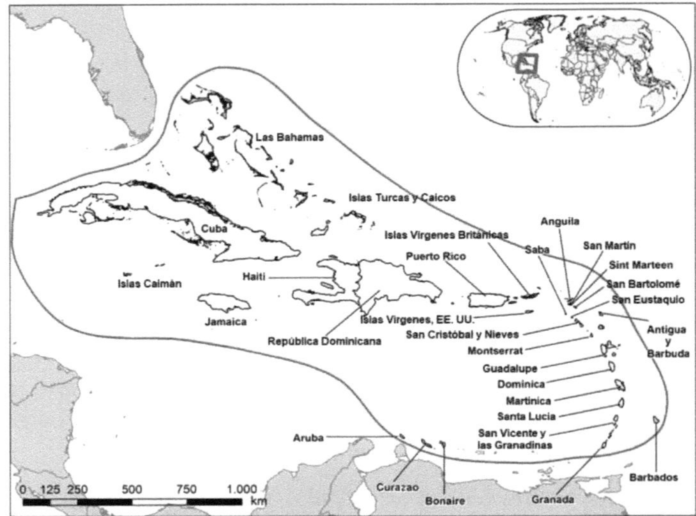

Figure 1. Biodiversity Hotspot of the West Indies

The West Indies hotspot comprises more than 7,000 islands, islets, islets, reefs and cays with a land area of 230,000 km^2 distributed over 4 million km^2 of sea (Figure 1.1). The hotspot encompasses 30 biologically and culturally diverse nations and territories from the following island groups: The Bahamas, the Greater Antilles, the Virgin Islands, the Cayman Islands, the Lesser Antilles, the Netherlands Antilles and the island group of Trinidad & Tobago. This represents a complex region of 13 independent countries and several overseas countries, territories and outermost regions of the Netherlands, France, the United Kingdom and the United States (Table 1.1). This group of islands hosts populations of endemic plants and vertebrates, which represent at least 2 % of the world's total species. Species endemism is very high within the region, despite the fact that the land area of the insular Caribbean is only 230 000 km^2 (90% of which corresponds to Cuba, Hispaniola, Jamaica and Puerto Rico).

Table 1.1 Countries and territories of the Antilles hot spot

Tabla 1.1 Países y territorios del hotspot de las islas del Caribe

Estados independientes	Territorios y países de ultramar			
	Francia	Reino de los Países Bajos	Reino Unido	Estados Unidos
Antigua y Barbuda*	Guadalupe	Aruba	Anguila	Navassa
Mancomunidad de las Bahamas*	Martinica	Bonaire	Islas Vírgenes Británicas	Puerto Rico
Barbados*	San Bartolomé	Curazao	Islas Caimán	Islas Vírgenes de Estados Unidos
Cuba	San Martín	Saba	Montserrat	
Mancomunidad de Dominica*		San Eustaquio	Islas Turcas y Caicos	
República Dominicana*		Sint Maarten		
Granada*				
Haití*				
Jamaica*				
Santa Lucía*				
San Cristóbal y Nieves*				
San Vicente y las Granadinas*				

Notas: * = países elegibles del CEPF.

2. BIOLOGICAL IMPORTANCE OF THE HOT SPOT
OF Biodiversity of the Antilles

Biodiversity hotspots are terrestrial regions that have at least 1,500 vascular plant species confined to them and have lost at least 70% of their original natural habitat (Mittermeier *et al.* 2004). The West Indies biodiversity hotspot is one of 36 areas in the world that meet these criteria. It is one of the world's major centers of endemic biodiversity, a result of the region's geography and climate: an archipelago of tropical and semi-tropical islands rich in habitats and tenuously connected to the surrounding mainland.

The Antilles hotspot consists mainly of three major island groups between North and South America: The Bahamas; the Greater Antilles; and the Lesser Antilles. It has an extremely complex geological history. Dispersal processes from North, Central and South America, Africa and Europe, climatic events and *in situ* radiations from the islands themselves, which are not yet fully understood, have resulted in exceptional plant diversity (WWF and IUCN 1997; Caujapé-Castells 2011; Nieto -Blázquez *et al.* 2017). There are 11 000 plant species, of which almost 8 000 are endemic (Acevedo-Rodríguez and Strong, 2008).

The biotas of these islands share an "oceanic" character (with the notable exception of the Trinidad & Tobago island group which is continental in character), marked by a relatively low representation of higher taxa, but there is extraordinary diversity within the higher phyletic groups present. Vertebrate diversity and endemism in the hotspot are also notable (Mittermeier *et al.* 2004). As a result of the high proportion of plants and animals endemic to the region, the West Indies hotspot is considered among the top five hotspots in the world (Myers *et al.*, 2000; Mittermeier *et al.*, 2004; Smith, 2004).

This chapter describes the importance of the Antillean hotspot from a geographical, geological, climatological, biogeographical, biological and ecological perspective. It also describes the importance of the hotspot in terms of the ecosystem services it provides to the human population.

2.1 Geography and climate

The West Indies hotspot is located on the Caribbean plate (except Cuba and Bahamas, which belong to the North American tectonic plate) and comprises more than 7,000 islands, islets, reefs and cays with a land area of 230,000 km^2 distributed over 4 million km^2 of sea. The island arcs delineate the eastern and northern edges of the Caribbean Sea: a semi-enclosed basin of the western Atlantic Ocean, with an area of approximately 2.75 million km^2 between Florida to the north and Venezuela to the south. Southeast of the Gulf of Mexico. The islands form a barrier between the Caribbean Sea and the Atlantic Ocean and can be divided into four main groups:

- The Greater Antilles (Cuba, Hispaniola, Jamaica, Puerto Rico, and the Virgin and Cayman Islands) represent approximately 90% of the land area of the hotspot. They are located on a partially elevated shelf that supports a mature volcanic mountain range and form the northern and western boundary of the Caribbean Sea.
- The Lesser Antilles are of more recent origin. They consist of an outer chain of low-lying coral and limestone islands and an inner chain of rugged volcanic islands on the eastern edge of the Caribbean Sea. The Leeward and Windward Islands extend from Anguilla in the north to Grenada in the south. Aruba, Bonaire and Curaçao border the southern end of the Caribbean Sea.
- The Bahamas Bank assemblage (including the Turks and Caicos Islands) rises from an underwater rock plateau southeast of Florida. Geographically, these islands are located in the Atlantic Ocean north of Cuba, not in the Caribbean Sea.
- The unique fauna assemblages of the insular group known as Trinidad & Tobago, the only continental islands of the Antilles archipelago, located in front of the Orinoco River delta, in the Gulf of Paria, adjacent to the northern coast of Venezuela. It has the characteristic fauna of this jungle region with the presence of monkeys, sloths, felines, tapirs, peccaries,

macaws, toucans and flamingos typical of the forest on the mainland.

Some islands in the hotspot have relatively flat terrain of non-volcanic origin. These islands include Aruba (which has only minor volcanic features), Barbados, Bonaire, the Cayman Islands, and Antigua. Others, such as Cuba, Dominica, Grenada, Grenada, Guadeloupe, Hispaniola, Jamaica, Montserrat, Puerto Rico, St. Lucia, and St. Vincent, have steep, elevated mountain ranges.

The highest mountain ranges rise to more than 3,000 m above sea level (Dominican Republic), while low-lying islands such as Anguilla, the Bahamas and the Turks and Caicos Islands reach just over 5060 m above sea level.

The climate in the West Indies is regulated by the north and southeast trade winds that meet at the intertropical convergence zone and two main western Atlantic currents (north and south equatorial currents) that converge to form the Caribbean Current: a warm current that transports significant amounts of the water northwestward through the Caribbean Sea and into the Gulf of Mexico via the Yucatan Current (Miloslavich *et al.* 2010, Gyory *et al.* 2018).

The Caribbean climate is humid tropical, but locally the climate and rainfall vary with elevation, island size and ocean currents (e.g., cool upwelling keeps Aruba, Bonaire and Curaçao semi-arid). It is moderated, to some extent, by warm, moist trade winds blowing steadily from the northeast, creating moist/semi-desert forest divisions on the mountainous islands.

At sea level, there is little variation in temperature, regardless of the time of day or season, with a range of 24 to 32°C (75 to 90°F). Rainfall distribution is determined by the size, topography and position of the islands in relation to the trade winds. Flat islands receive slightly less rain, although rainfall is more predictable. The heaviest rainfall periods occur in mid-May and September (although with a temporal variation in the hot spot), with the "rainy season" coinciding with the summer hurricane season.

Hurricanes develop over the ocean during the middle and later months of the year (June to November) when sea surface temperatures are high (above 27°C) and air pressure drops below 950 millibars.

Caribbean winters are warm but drier, although occasional northwesterly winds bring cooler conditions to the northern islands in the winter. Caribbean waters are mostly clear and warm (2229°C) and the tidal range is very low (<0.4 m) (Miloslavich *et al.* 2010).

2.2 Habitats and ecosystems

The geography, climate and large geographic extent of the Antillean hotspot have resulted in a wide range of habitats and ecosystems, which in turn support high levels of species richness.

Fourteen Holdridge life zones and 16 World Wildlife Fund (WWF) ecoregions have been defined in the hotspot. There are four main terrestrial forest types, whose distribution and biodiversity characteristics are described below:

Humid tropical/subtropical rainforests (pluvisilvas) occur mainly in lowlands influenced by northeast or northwest winds and on the windward slopes of mountains, such as the northern part of eastern Cuba, northern Jamaica, eastern Hispaniola, northern Puerto Rico and small patches in the Lesser Antilles.

Tropical/subtropical dry broadleaf forests are found in The Bahamas, Cayman Islands, Cuba, Hispaniola, Jamaica, Lesser Antilles and Puerto Rico.

The dry forest life zone tends to be favored for human occupation, largely because of the relatively productive soils and reasonably comfortable climate. For this reason, few dry forests remain intact.

Tropical/subtropical coniferous forests (lowland and montane) are found in The Bahamas, Turks and Caicos Islands, Cuba and Hispaniola, where they are often threatened by timber extraction and frequent human-induced fires, which change their age structure and density.

Xeric scrub and xeric thickets occur in rain shadow areas created by mountains, as well as in the more arid climate of the southern Caribbean (e.g., Aruba, Bonaire, and Curaçao). Xeric scrub and cactus scrub are found where conditions are suitable in the Lesser Antilles, Cuba and Hispaniola.

In the marine realm, the shallow marine environment of the Antilles is part of the large marine ecosystem of the Caribbean Sea, with more than 12,000 reported marine species and low rates of endemism compared to terrestrial ecosystems, due to the high degree of connectivity resulting from the influence of currents and species migration (Miloslavich *et al.* 2010).

The Caribbean coastal zone contains many productive and biologically complex ecosystems, including beaches, coral reefs, seagrass beds, mangroves, coastal lagoons, and mud bottom communities (UNEP-RCU, 2001). The health of these ecosystems has declined over the years, mainly due to habitat conversion, overexploitation, and pollution from suspended solids and chemical compounds (Polunin and Williams, 1999; AIMS, 2002; Lang, 2003).

The beaches are among the most important coastal ecosystems in the Caribbean. They provide important habitats for several species, including nesting sites for large sea turtles, and are of great economic importance for tourism in the region. They are dynamic environments, constantly changing due to natural processes such as storms, hurricanes, tidal changes and sea level rise.

Most corals and coral reef-associated species in the Caribbean Sea are endemic, making the region biogeographically unique (AIMS, 2002; Spalding, Green and Ravilious, 2001). In addition to the important ecosystem services they provide, coral reefs are of fundamental economic importance to the Caribbean, particularly for tourism and fisheries (Heileman, 2005).

Seagrass meadows are usually found in areas protected by coral reefs and comprise predominantly two species: turtle grass - *(Thalassia testudinum); and* manatee grass - *(Syringodium filiforme).* These productive habitats are grazing grounds for the green turtle - *(Chelonia mydas),* the West Indian manatee - *(Trichechus manatus) and* many other vertebrates and invertebrates; they also contribute to water clarity.

Coastal wetlands, including estuaries, coastal lagoons and other coastal marine waters, are highly fertile and productive ecosystems. Mangroves and littoral forests are considered the most biologically diverse marine habitats after coral reefs.

Mangroves and seagrass meadows serve as nurseries for juveniles of many commercially important fish species, while providing habitat for a wide variety of small fish, crabs and birds. They also play an important role in coastal protection against weather events in an area like the Caribbean that is affected by hurricanes every year.

Soft-bottom hotspot habitats are species-rich. Flaccid bottom habitats include environments where the seafloor consists of fine-grained sediments, mud and sand. Intertidal or shallow intertidal soft-bottom habitats include mudflats and seagrass meadows, which are economically and ecologically important. They are inhabited by burrowing animals, such as worms, snails, clams and some anemones, shrimps and crabs, sea crackers, brittle starfish and sea cucumbers. Several fish species feed in muddy soft bottom habitats (Halpern *et al.* 2008).

2.3 Species diversity and endemism

The hot spot of the Antilles supports a wealth of biodiversity within its diverse ecosystems, with a high proportion of endemicity, which makes the region biologically unique and therefore its fauna very special.

It includes about 11 000 plant species, 72 % of which are endemic (Acevedo-Rodriguez and Strong, 2007). In vertebrates, high proportions of endemic species characterize the herpetofauna (96 % of 200 amphibian species and 82 % of 602 reptile species), which is probably due to their low dispersal rates, in contrast to birds (26 % of 565 species) and mammals (49 % of 104 species, most of which are bats) (BirdLife International, 2017; IUCN, 2017a). Endemic species in the hotspot represent 2.5 % of the 310 442 plant species described worldwide and 1.4 % of the 68 574 vertebrate species described worldwide (IUCN, 2017a).

Data for marine species are still incomplete. The approximately 12 000 marine species recorded so far in the Caribbean is a clear underestimate for this diverse tropical region. Sampling efforts, to date, have been heavily biased towards certain habitats in shallow coastal waters, particularly coral reefs; very little information is available on benthic organisms deeper than 500 m (Miloslavich *et al.,* 2010).

2.3.1 Mammals

Historically, the West Indies hotspot was home to 127 species of terrestrial mammals, 23 of which are now considered extinct. Of the 104 extant species, 51 are endemic to the hotspot. Solenodontidae and Capromyidae are two families of rodents endemic to the Greater Antilles. The family Solenodontidae includes two surviving species, both Endangered (EN): the Cuban soledon *(Atopogale cubanus); and* the Hispaniolan solenodon *(Solenodon paradoxus).*

The Cuban almiquí occurs in two national parks: Alejandro de Humboldt; and Sierra del Cristal. In Haiti, the Hispaniolan solenodon is known to occur only in the Massif de la Hotte mountain range, but its distribution is more widespread in the Dominican Republic. The main threats are habitat loss due to increased human activity and deforestation, and the introduction of exotic predators such as dogs, cats and mongooses.

The rodent family Capromyidae (hutias) comprises 16 species, 15 of which are found in the hotspot. Five of these species are extinct due to hunting, habitat loss, and predation by invasive species.

The 10 remaining species are country-specific, with seven species in

Cuba and a single species in The Bahamas *(Geocapromys ingrahami - Vulnerable (VU))*, Jamaica *(G. brownie - VU)* and Hispaniola *(Plagiodontia aedium - EN)*. However, two of the Cuban endemic species are considered "possibly extinct", namely the dwarf hutia *(Mesocapromys nanus - CR)* and the land hutia (*M. sanfelipensis - CR*), while the rat hutia *(M. auritus - EN)* is restricted to a single site on the Cuban islet of Cayo Fragoso.

Bats are very important components of the ecosystems in the West Indies hotspot and are represented by 59 species (except in the sub-archipelago of Trinidad & Tobago, with more than 100). However, there is an urgent need to study bats to better understand their distribution, ecology and current threat status. These species are sparsely distributed and difficult to find due to the limited number of suitable caves or suitable mature (native) roosting trees. For example, the big funnel-eared bat *(Natalus primus - CR)* is only found in La Barca Cave in Guanahacabibes in far western Cuba; while the Jamaican funnel-eared bat *(Natalus jamaicensis - CR)* was only recorded in St. Clair Cave in the Point Hill CBA and Portland Cave in the Portland Bight Protected Area CBA.

In the insular group of Trinidad & Tobago there are primates, sloths, tapirs, peccaries and felids typical of the jungles of the Orinoco River, typical of the fauna found in the mainland forest.

2.3.2 Birds

A total of 571 bird species have been recorded in the West Indies hotspot (BirdLife International, 2017), six of which are already extinct. Of the 565 extant species, 147 are endemic to the hotspot, 105 of which are confined to individual islands. Although endemism is most notable at the species level, 36 notable genera of birds are endemic to the hotspot, as well as two endemic families: Dulidae (palm cigua, *Dulus dominicus),* with one species; and Todidae (barrancolíes, cortabubas, with five species). The Caribbean is also home to the world's smallest bird, the hummingbird *(Mellisuga helenae).*

BirdLife International recognizes six endemic bird areas (EBAs) and two secondary areas within the West Indies hotspot (Stattersfield *et al.,* 1998), a testament to the diversity and endemism (linked to a particular island) in this region.

Birds represent some of the most important symbols for conservation in the Caribbean. Parrots, including the St. Vincent amazon *(Amazona guildingii* - VU), St. Lucia amazon *(A. versicolor* - VU) and Dominica's imperial amazon (*A. imperialis* - EN), have been successfully used as flagship species for conservation and environmental awareness in their respective countries.

2.3.3 Reptiles

With more than 600 native species, the Antilles are very rich in reptiles, the vast majority of which (about 82 %) are endemic to the region.

Since the last CEPF ecosystem profile was published in 2010, at least 39 new species have been described, including several lizards and anoles and two boas (Hedges and Conn, 2012; Kolher and Hedges,

2016; Reynolds *et al.* 2016; Hedges, 2018). Many of the hotspot species are endemic to a particular island and may be extinct or near extinction. These new species have not yet been formally assessed according to IUCN Red List criteria, and other taxa are still in the process of being formally accepted as valid new species (Morton, 2009).

Two main evolutionary radiations dominate the lizards: anoles (genus *Anolis,* 166 species) and dwarf geckos (genus *Sphaerodactylus,* 85 species). Notable reptile taxa also include 11 species of rock iguanas *(Cyclura spp.),* 10 of which are globally threatened, and the poorly known and elusive pikes (27 species in two genera, *Celestus* and *Diploglossus'),* some of which are feared extinct. Two of the world's smallest lizards are found in the *Caribbean: Sphaerodactylus ariasae* from the Dominican Republic; and *S. parthenopion* (EN) from the U.S. Virgin Islands.

Snakes comprise 149 native species in nine families, and include important radiations such as the genus *Tropidophis* (27 species), a group of dwarf boas and the genus *Typhlops* (41 species) of blind snakes. The world's smallest snake, the Barbados thread snake (*Tetracheilostoma carlae* - CR), is known to occur in only a very small area of Barbados (Hedges 2008).

Four species of sea turtles nest in the Caribbean: the leatherback (*Dermochelys coriacea*); hawksbill (*Eretmochelys imbricata*); green (*Chelonia mydas); and* loggerhead *(Caretta caretta);* all are globally threatened. Some authors have estimated that the historical populations of these species in the Caribbean numbered in the millions (Jackson, 1997). So abundant were they that seafarers' reports from the 17th and 18th centuries document flotillas of turtles so dense and vast that it was impossible to fish them with nets, and they even impeded the passage of ships (Harold and Eckert, 2005; WIDECAST, 2018).

Today, sea turtle populations have declined considerably compared to historical levels, and some of the largest breeding populations have disappeared (Harold and Eckert, 2005).

2.3.4 Amphibians

All 200 native amphibian species in the Caribbean are endemic, many are endemic to particular islands (IUCN, 2017b). This number is likely to increase as more research is conducted in more remote areas of the region, particularly in the Greater Antilles.

Amphibians belong to six families of frogs (Aromobatidae, Bufonidae, Craugastoridae, Eleutherodactylidae, Hylidae, and Leptodactylidae), but the taxon is dominated by the 152 species of the genus *Eleutherodactylus.* These forest frogs are distinctive because of their direct development (no metamorphosis or bypassing the tadpole stage), egg-laying on the ground, and parental protection of the eggs. *Eleutherodactylus iberia* from Cuba is one of the smallest tetrapods in the world, measuring less than 1 cm in length.

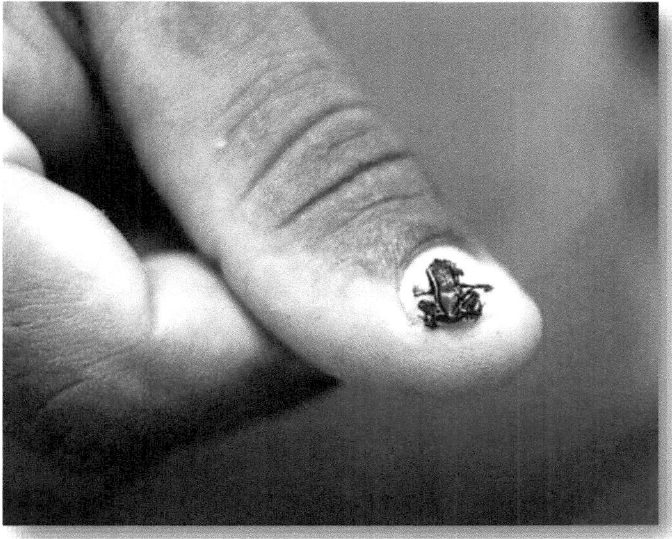

At the other end of the scale, the mountain chicken *(Leptodactylus fallax)* of Montserrat and Dominica measures 16 cm and is one of the largest frogs.

This species is one of many amphibians that have fallen victim to an infectious disease caused by the chytrid fungus *Batrachochytrium dendrobatidis,* compounded by historical impacts of habitat loss, invasive species and exploitation, is rapidly declining towards extinction in the wild on both islands being one of the fastest declines ever recorded affecting the entire range of a species (Hudson *et al.,* 2016).

Recent efforts have succeeded in bringing the mountain chicken back from the brink of extinction, but its status remains perilous. The disease has also been implicated in the rapid declines and possible extinctions of several species of *Eleutherodactylus* in Puerto Rico, the Dominican Republic, Haiti and Cuba.

Along with disease, amphibians in the region face threats from invasive species, as well as habitat loss and fragmentation.

2.3.5 Freshwater fishes and coastal marine fishes

IUCN has assessed the global threat status of a total of 1 538 bony fish species in the West Indies hotspot; this represents about 4 % of all bony fish species.

The species list for the hotspot is still incomplete and new species are being discovered on mesophotic and deep reefs (Baldwin and Robertson, 2014; Baldwin and Robertson, 2015; Baldwin *et al.,* 2016 a and b; Tornabene *et al.,* 2016).

The West Indies hotspot contains 167 species of freshwater fish, approximately 65 of which are endemic to one or a few islands, and many of which are found only in a particular lake or spring.

As in other hot spots of the archipelago, there are two distinct groups of freshwater fish in the Caribbean: on the smaller, younger islands, most fish species are marine water but also enter freshwater to some extent, while on the larger, older islands of the Greater Antilles, there are several groups that occupy inland waters, such as needlefish, killifish, silversides and cichlids, such as the mosquitofish *(Gambusia dominicensis* - EN), which is restricted to lakes Enriquillo and Azuéi (both CBAs).

Marine fishes represent a complex group of organisms, including many important fishery species, such as the American eel *(Anguilla rostrata* - EN) and several species of groupers, including the Nassau grouper *(Epinephelus striatus* - EN).

The Wider Caribbean biogeographic region (includes areas outside the hotspot, such as Bermuda, the Gulf of Mexico, and Trinidad and Tobago) contains the highest marine species richness in the Atlantic Ocean and is considered a global hotspot for tropical reef species (Roberts *et al.,* 2002).

A study of the conservation status of coastal bony fishes in the Wider Caribbean found that 53 % of the 1 360 species included in the study were endemic, with the highest degree of endemism being in the Atlantic Ocean (Linardich *et al.,* 2017).

Offshore oceanic areas have lower species richness, due to the resource-poor environment and limited opportunity for niche di versification. However, most endemic species tend to be widely distributed, probably due to the high level of marine connectivity in the region.

2.3.6 Sharks

There are 83 species of chondrichthyans (cartilaginous fishes) in the marine waters of the Antillean hotspot. However, only 59 of these are found in nearshore waters (deeper than 200 m). These species are from 27 families, comprising 16 families of sharks (44 species) and 11 families of rays (15 species).

Most species have extensive ranges and some are found worldwide. However, there appears to be at least one endemic species, the Florida torpedo *(Torpedo andersoni),* which is known from only two specimens: one from the western edge of the Grand Banks of the Bahamas; and the other from a coral reef in Grand Cayman.

2.3.7 Reef-building corals

Coral reefs are among the most important coastal marine ecosystems in the hotspot and play a critical role in the region's economy. The livelihoods of millions of people depend on the reefs for income and employment.

In the Caribbean Sea, corals represent a biogeographically distinct area within which most corals and their associated species are endemic, making the entire region particularly important in terms of global biodiversity (Spalding *et al.,* 2001; AIMS, 2002).

Caribbean coral reefs include more than 65 reef-building species; many of these are widely distributed, but are also endemic to the region due to the long isolation of the western Atlantic from the Pacific. Among the most widespread genera are *Acropora, Monastrea, Porites, Agaricia, Diploria, Colpophylia, Meandrina, Mycetophyllia, Dendrogyra and Millepora.*

The area covered by coral reefs in the Caribbean has been estimated at 26 000 km^2 , or approximately 10 % of the total worldwide (Keith *et al.,* 2013).

2.5 Globally threatened species

The total land area of the West Indies hotspot is only 230 000 km^2. With only 10% of the original hotspot habitat remaining, most of the major habitat loss has already occurred. However, with population growth (albeit slowing) and changes in land use patterns, what little habitat remains is at risk from both human activity and natural disasters.

Table 4.1: Species diversity, endemism and global threat status in the Antillean hotspot.

Taxonomic group	Species	Species endemic to the point hot	Percent endemic	Threatened species at the world	Percent threatened
Mammals	104,0	51,0	49,0	26,0	25,0
Birds	565,0	148,0	26,2	55,0	9,7
Reptiles	602,0	494,0	82,1	184,0	30,6
Amphibians	200,0	191,0	95,5	146,0	73,0
Bony fish	1,538	65,0	4,2	42,0	2,7
Cartilaginous fish	83,0	-	-	17,0	20,5
Reef-building corals	91,0	-	-	15,0	16,5
Plants with seeds	10 948	7 868	71,9	507	4,6

Total	14 134	8 817	62,4	992	7,0

2.7 Ecosystem services

The Millennium Ecosystem Assessment (2005) grouped ecosystem services into four categories: provisioning services such as food, water, timber and fiber; regulating services that moderate climate, flooding, disease and water quality; cultural services that provide recreational, aesthetic and spiritual benefits; and supporting services such as soil formation, photosynthesis and nutrient cycling.

Although some studies on ecosystem services have been conducted in the Antilles, there is much less information available on the ecosystem and ecological services of the hotspot than in other regions of the Americas.

The available information is fragmented and has not yet been compiled at the hotspot scale. Some authors and organizations have assessed and valued ecosystem services at the ecosystem level, and in some rare cases at the country level. Others have focused on ecosystem services in protected areas.

Bovarnick *et al.,* (2010) assessed the importance of biodiversity in the region and other authors (e.g. Heileman 2005) provided general information on ecosystem services in the region. Mumby and Fitzsimmons (2014) reviewed the ecosystem services of reefs in the Caribbean. John (2005) assessed the contribution of non-timber forests to rural economies in the Caribbean Windward Islands.

At the country level, a cost analysis of ecosystem services provided by the National System of Protected Areas was carried out in the Dominican Republic (Gómez-Valenzuela *et al.,* 2014).

Assessments of water-related services were conducted in Jamaica (Pantin and Reid, 2005) and St. Lucia (Springer, 2005).

Greenhouse gas sequestration was assessed in Montserrat (Peh *in litt.) and* carbon sequestration in southwest Tobago (Varty, 2016).

Cesar *et al.* (2000) and Guingand (2008) assessed fisheries, forestry (charcoal and other non-timber products), tourism, recreation, waste treatment, sediment retention, coastal protection, carbon sequestration, biodiversity, and cultural heritage at the Portland Bight Protected Area site in Jamaica. Edwards (2011 and 2013) assessed the economic value of carbon sequestration in the Cockpit Country and Coral Spring and Mountain Spring Protected Area in Jamaica.

CBAs in the West Indies hotspot are important for their species richness and uniqueness, but they are also extremely important sources of provisioning, regulation and cultural ecosystem services.

All hotspot ecosystems and, by extension, many of its CBAs, provide multiple ecosystem services. Forests, for example, are important for erosion control, flood mitigation, water purification, pollination, waste assimilation and disease regulation.

Table 4.2 provides a summary of the main ecosystem services in the hotspot and the following sections provide a discussion of some ecosystem services of major importance to the hotspot countries.

Table 4.2 Main ecosystem services in the Antillean biodiversity hotspot.

Ecosystem service	Beneficiaries	Relative importance in the hot spot
Procurement services		
Freshwater flows (artesian and runoff, flows) for consumption, irrigation, industrial use, energy generation, etc.	The entire population of the hot spot	Very important as the area is under water stress.
Food production (Fishing in freshwater and marine systems)	Local fishermen, fish consumers, associated economic activities	Very important for local hotspot fishing communities
Food production (Crops)	The entire population of the hot spot	Very important
Food production (Livestock)	The entire population of the hot spot	Very important
Non-timber forest products. (Honey, handicraft materials, straw, ornamental and household plants, spices, oils, seeds, seedlings, orchids, fruits)	The entire population of the hot spot	Important, some products may be obtained from other sources outside the hot spot.
Wood products	The entire population of the hot spot	Very important
Medicinal and pharmaceutical plants.	Rural communities	Very important
Energy (Solar and wind energy)	The entire population	Important
Support Services		
Habitat for species	World Cup	Very important for global biodiversity

Maintenance of genetic diversity (Source of novel genetic material for crops (e.g. fruits).	World Cup	Potentially significant

Criterion A: Threatened biodiversity.
Criterion B: Geographically restricted biodiversity.
Criterion C: Ecological integrity.

3. CONSERVATION OUTCOMES DEFINED FOR THE ANTILLES BIODIVERSITY HOTSPOT

Biological diversity cannot be saved through *ad hoc* actions (Pressey, 1994). To support the implementation of coordinated conservation actions, CEPF invests in defining conservation outcomes to identify a quantifiable set of species, sites, and corridors that must be conserved to promote the long-term persistence of the world's biodiversity.

By presenting quantitative, defensible, and verifiable targets against which to measure the success of investments, conservation outcomes allow the limited resources available for conservation to be more effectively identified and impacts monitored on a global scale. Conservation outcomes are the basis for identifying biological priorities for CEPF investment in the West Indies hotspot.

Biodiversity is not measured by a single unit, but is distributed across a hierarchical spectrum of ecological scales (Wilson, 1992). This spectrum can be reduced to three levels: species, sites and corridors (landscapes of interconnected sites). These three levels are geographically intertwined, through the occurrence of species in sites, and species and sites in corridors.

Given the threats to biodiversity at each of the three levels, quantifiable targets for conservation can be set in terms of avoided extinctions (species outcomes), protected areas (site outcomes), and consolidated corridors (corridor outcomes). Conservation outcomes are defined sequentially, with species outcomes defined first, then site outcomes, and finally corridor outcomes.

CEPF defines species outcomes as globally avoided extinctions, which are directly linked to globally threatened species using the IUCN Red List categories of Critically Endangered, Endangered and Vulnerable. This definition excludes Data Deficient species, which are considered a priority for future research, but not necessarily for conservation action per se.

Species outcomes are achieved when the global threat status of a species improves or, ideally, when it is removed from the Red List.

The basis for defining the species scores for the West Indies biodiversity hotspot profile is the global threat assessments contained in the *IUCN Red List of Threatened Species (2017), the* authoritative data source for the global conservation status of species.

Since most globally threatened species in the Caribbean are best conserved by protecting a network of sites where they occur, the basis for defining site scores is the full suite of CBAs in the hotspot.

CBAs are sites of importance for the global persistence of biodiversity. They are identified by the elements of biodiversity that specific sites contribute significantly to global persistence, such as globally threatened species or ecosystems.

The identification of CBAs follows the *Global Standard for the Identification of Key Biodiversity Areas,* prepared by the IUCN Species Survival Commission and the IUCN World Commission on Protected Areas in partnership with the IUCN Global Species Programme (IUCN, 2016).

The CBA Standard includes a total of five criteria and 11 sub-criteria under which a site can be identified as a CBA. For this ecosystem profile, only seven of the 11 sub-criteria were used to identify CBAs in the West Indies: threatened species (Criteria A1a-e) for all Critically Endangered, Endangered and Vulnerable species, individually geographically restricted species (B1) and demographic aggregations (D1, for some birds only).

Site outcomes are achieved when a CBA is protected through improved management, expansion of an existing conservation area or creation of a new conservation area.

Improved management of an existing conservation area involves changing the management practices of the CBA to improve the long-term conservation of species populations and the ecosystem as a whole.

The process of expanding an existing conservation area involves increasing the proportion of a CBA under conservation management to meet species area requirements or to incorporate other previously excluded species or habitats.

The creation of a new conservation area involves designating all or part of a CBA as a conservation area and initiating effective long-term management.

Conservation areas are not limited to actual or potential protected areas, but also include what has been defined as 'Other Effective Area-based Conservation Measure' (OECM) (Jonas *et al.*, 2014), which includes sites managed for conservation by local communities and private landowners or other stakeholders.

The update and identification of CBAs in the Caribbean according to the CBA Standard, 2016; took into account sites identified as CBAs in the 2009 CEPF ecosystem profile based on the previous CBA standard (Langhammer *et al.*, 2007), the 2017 AZE update and new protected areas declared since 2009.

This update was conducted through analysis of regionally accessible data (databases, museum specimens, etc.) and literature reviews, with support from IUCN and the New York Botanical Garden, followed by consultations with local experts in the Dominican Republic, Haiti, and Jamaica, and an online workshop with experts from The Bahamas and Lesser Antilles. Information for Cuba and the U.S. and European territories is based on previous assessments and was not reviewed or updated for this ecosystem profile (Brown *et al.*, 2019).

While protection of a network of sites may be sufficient to conserve most elements of West Indian fauna in the medium term, long-term conservation of biodiversity often requires the consolidation of landscapes of interconnected sites or "conservation corridors", especially in landscapes on the larger islands. This is particularly important for the conservation of large-scale ecological and evolutionary processes (Schwartz, 1999), and to ensure ecosystem resilience.

In order to allow for the persistence of biodiversity, landscapes with interconnected sites must be anchored in core areas embedded in a matrix of natural and/or anthropogenic habitats (Soulé and Terborgh, 1999). Therefore, conservation corridors are anchored in CBAs, with the remainder of the conservation corridor comprising areas with the potential to become CBAs in their own right (through management or restoration) or areas that contribute to the conservation corridor's ability to support all elements of biodiversity over the long term.

The CBAs, therefore, were the starting point for defining conservation corridors in the Antilles hotspot, especially in the larger islands.

First, conservation corridors were defined where it was necessary to maintain connectivity between two or more CBAs to meet long-term biodiversity conservation needs. Then, additional conservation corridors were defined where it was deemed necessary to increase the area of actual or potential natural habitat to maintain evolutionary and ecological processes. In the latter case, the definition of conservation corridors was largely subjective, due to time constraints, lack of relevant data and absence of detailed criteria.

Given these limitations, emphasis was placed on maintaining natural habitat continuity across environmental gradients, particularly altitudinal gradients, in order to maintain ecological processes such as altitudinal migration of bird species and to provide protection against the potential impacts of climate change.

Conservation corridors were defined through consultation with local experts, supplemented by analysis of other data layers and published reports (Brown, *ob. cit.*, 2019).

Due to the fragmented nature of an archipelagic hotspot like the West Indies (often with isolated CBAs established in developed or highly degraded landscapes), defining landscape-scale outcomes was not always appropriate. As a result, relatively few conservation corridors were defined overall.

In theory, within a given region or, ultimately, for the entire world, conservation outcomes can be defined for all taxonomic groups. However, this depends on the availability of data on the global threat status of all taxa and the distribution of globally threatened species among sites and corridors.

In the West Indies hotspot, data were only available for mammals, birds, amphibians, reptiles, fish and corals, so conservation results were only defined for these groups.

The approach of using global threat assessments as the basis for defining species outcomes, and consequently site and corridor outcomes, has several limitations, the most serious being that these assessments are incomplete for many taxonomic groups. In addition, defining conservation outcomes is an iterative process: as more species are assessed as globally threatened, additional site outcomes may be defined. As irreplaceability criteria are applied to taxa other than birds, these additional site-level outcomes can help fill gaps in taxonomic coverage.

3.1 Species results

The fauna of the Antillean hotspot is at serious risk of species extinction. Out of a total of more than 14 000 species (of the taxa included in the assessment, Table 4.1), 4 182 species have been assessed using the IUCN Red List criteria, of which 992 species (24 %) are globally threatened.

A complete list of globally threatened taxa used for this ecosystem profile is available in Appendix 1 of Brown *et al.,* (2019); and a summary of the total number of threatened species in each hotspot country can be found in Table 5.1. The hotspot is particularly important for reptiles and amphibians, due to high rates of speciation and endemism and exceptionally high levels of threat.

The following sections characterize the CEPF eligible countries and the main focus of this ecosystem profile update. They also highlight areas where new information has been provided.

Rather than detailing the number of threatened and endemic species by country, this section presents a general discussion of species that trigger CBA criteria for one or more sites. These are the species that CEPF will focus its efforts on when aligning priority species in priority sites (CBAs).

Table 5.1 Summary of threatened species by country - West Indies Biodiversity Hotspot

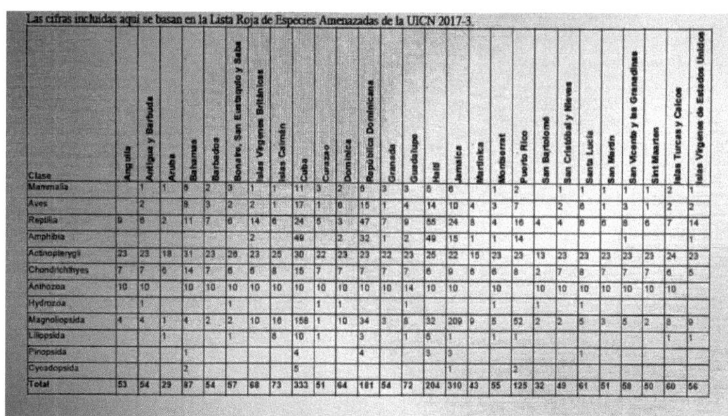

Las cifras incluidas aquí se basan en la Lista Roja de Especies Amenazadas de la UICN 2017-3.

Clase	Anguila	Antigua y Barbuda	Aruba	Bahamas	Barbados	Bonaire, San Eustaquio y Saba	Islas Vírgenes Británicas	Islas Caimán	Cuba	Curazao	Dominica	República Dominicana	Granada
Mammalia		1	1	6	2	3	1	1	11	3	2	6	3
Aves		2		9	3	2	2	1	17	1	8	15	1
Reptilia	9	8	2	11	7	6	14	6	24	6	3	47	7
Amphibia						2			49		2	32	1
Actinopterygii	23	23	18	31	23	26	23	25	30	22	23	23	22
Chondrichthyes	7	7	6	14	7	6	6	8	15	7	7	7	7
Anthozoa	10	10		10	10	10	10	10	10	10	10	10	10
Hydrozoa		1				1				1	1		1
Magnoliopsida	4	4	1	4	2	2	10	16	158	1	10	34	3
Liliopsida			1			1		8	10	1		3	
Pinopsida				1					4			4	
Cycadopsida			2						5				
Total	53	54	29	87	54	57	68	73	333	51	64	181	54

Clase	Guadalupe	Haití	Jamaica	Martinica	Montserrat	Puerto Rico	San Bartolomé	San Cristóbal y Nieves	Santa Lucía	San Martín	San Vicente y las Granadinas	Sint Maarten	Islas Turcas y Caicos	Islas Vírgenes de Estados Unidos
Mammalia	3	5	6		1	2		1	1	1	1	1	2	1
Aves	4	14	10	4	3	7		2	6	1	3	1	2	2
Reptilia	9	55	24	8	4	16	4	4	6	6	8	6	7	14
Amphibia	2	49	15	1	1	14				1				1
Actinopterygii	23	25	22	15	23	23	13	23	23	23	23	23	24	23
Chondrichthyes	7	6	9	6	6	8	2	7	8	7	7	7	6	6
Anthozoa	14	10	10		10		10	10	10	10	10	10	10	
Hydrozoa				1			1		1					
Magnoliopsida	8	32	209	9	5	52	2	2	5	3	5	2	8	9
Liliopsida	1	5	1		1	1							1	1
Pinopsida		3	3			1								
Cycadopsida						1				2				
Total	72	204	310	43	55	125	32	49	61	51	58	50	60	56

Section 5.2 presents information on the different datasets used in this profile. More detailed information on the species results for Cuba, Puerto Rico and the U.S. Virgin Islands can be found in the ACB Global Database (http://www.keybiodiversityareas.org/site/results?reg=4andcty=53an dsnm).

Details of the species results for the EU OCTs can be found in the ecosystem profile prepared by the EU BEST initiative (Voluntary Scheme for Biodiversity and Ecosystem Services in the EU Outermost Regions and Overseas Countries and Territories) (http://ec.europa.eu/environment/nature/biodiversity/best/regions/cari bbean_e n.htm).

Of the 992 globally threatened species in the West Indies hotspot, 575 species are found in countries eligible for CEPF funding. Of these species, only 337 trigger CBA criteria based on information available at the time of the ecosystem profiling process (Table 5.2).

Eligible CEPF countries were Antigua and Barbuda, Bahamas, Barbados, Barbados, Dominica, Dominican Republic, Grenada, Haiti, Jamaica, St. Kitts and Nevis, St. Lucia, and St. Vincent and the Grenadines.

Table 5.2. Species results for CEPF eligible countries in the West Indies hotspot.

Class	CR	EN	VU	Total	Percentage of endangered species activating CBAs

Mammalia	2 (2)	4 (3)	8 (6)	14 (11)	78,6
Birds	7 (3)	11 (10)	19 (15)	37 (28)	75,7
Reptilia	44 (21)	50 (18)	24 (13)	118 (52)	44,1
Amphibia	40 (29)	30 (24)	8 (8)	78 (61)	78,2
Actinopterygii	3 (0)	6 (4)	24 (1)	33 (5)	15,2
Chondrichthyes	1 (1)	3 (0)	12 (0)	16 (1)	6,3
Anthozoa	2 (0)	2 (0)	6 (0)	10 (0)	0,0
Hydrozoa	0 (0)	1 (0)	0 (0)	1 (0)	0,0
Total	99 (56)	107 (59)	101 (42)	307 (158)	58,6

Eligible CEPF countries were Antigua and Barbuda, Bahamas, Barbados, Barbados, Dominica, Dominican Republic, Grenada, Haiti, Jamaica, St. Kitts and Nevis, St. Lucia, and St. Vincent and the Grenadines.

3.1.1 Mammals

Mammals constitute the taxonomic group with the lowest number of threatened species in the hotspot. There are 26 globally threatened mammal species, 14 of which occur in eligible countries. Eleven of these species trigger CBA criteria for one or more sites: two Critically Endangered species (*Phyllonycteris aphylla* bat *and* Jamaican funnel-eared bat (Natalus jamaicensis)), three Endangered species (Hispaniolan hutia, Hispaniolan solenodon and Guadalupe big-eyed bat *(Chiroderma improvisum))* and six Vulnerable species.

Three mammal species have not been reported in any CBA: the sei whale *(Balaenoptera borealis), the* blue whale *(Balaenoptera musculus) and* the Dominican myotis *(Myotis dominicensis).*

The identification of CBAs in the marine environment (beyond coastal areas) was outside the scope of this profile.

Four mammal species that trigger the CBA criteria (all bats) have each been confirmed for one site only. Two species have been reported from two sites. The remainder occur at four or more sites, the Jamaican hutia, Hispaniolan hutia, Hispaniolan solenodon, funnel-eared bat, and Jamaican red bat *(Lasiurus degelidus)* all occur at more than 10 sites (Table 5.3).

Table 5.3 Globally threatened mammals by country and number of CBAs (CEPF eligible countries only).

Familia	Especie	Lista Roja de la UICN	Número de ACB/país					
			República Dominicana	Bahamas	Barbados	Haití	Jamaica	San Cristóbal y Nieves
Capromyidae	Jutía jamaiquina (*Geocapromys brownii*)	VU					17	
	Jutía de Bahamas (*Geocapromys ingrahami*)	VU		2				
	Jutía de La Española (*Plagiodontia aedium*)	EN	15			2		
Natalidae	Murciélago orejas de embudo (*Natalus jamaicensis*)	CR					1	
Phyllostomidae	*Chiroderma improvisum*	EN						1
	Phyllonycteris aphylla	CR					1	
Solenodontidae	Solenodonte de La Española (*Solenodon paradoxus*)	EN	13			1		
Trichechidae	Manatí americano (*Trichechus manatus*)	VU	2			2	2	
Vespertilionidae	Murciélago rojo jamaiquino (*Lasiurus degelidus*)	VU					3	
	Lasiurus minor	VU	6			2		
	Myotis nyctor	VU			1			

3.1.2 Birds

With a total of 55 threatened species, birds rank third among the animals in the hotspot in terms of the number of threatened species. Thirty-seven threatened bird species occur in CEPF-eligible countries, four of which were not reported in any CBA. While some species may not trigger CBA criteria in the hotspot due to their marginal populations in the Caribbean, such as the Cerulean Warbler *(Setophaga cerulea - VU)* and Leach's Petrel *(Hydrobates leucorhous - VU)*, others cannot trigger CBA criteria as they are potentially globally extinct, such as the Eskimo curlew *(Numenius borealis - CR)*, or locally extinct, such as the Cuban tyrant *(Tyrannus cubensis - EN)*, which is no longer found in The Bahamas.

Of the 33 bird species reported in the CBAs, 28 trigger CBA criteria: three Critically Endangered, 10 Endangered and 15 Vulnerable (Table 5.4). Of particular relevance is the lack of sites triggering CBA criteria for the Jamaican nightjar (*Siphonorhis americana* - CR), Semper's warbler *(Leucopeza semperi* - CR) and Bahamas nuthatch (EN). These are country endemics in Jamaica, St. Lucia and Bahamas, respectively, and all lack recent records. Only one individual of the Bahamas nuthatch, originally reported for a single site (mature pine forests in Grand Bahama), has been recorded since 2016. Therefore, it could not be confirmed as a species triggering the CBA criteria at that site.

Despite their mobility, populations of some bird species are actually restricted to very few sites, such as the Ridgway's falcon (CR) or the imperial amazon (EN), both of which are known to occur at only two sites in the Dominican Republic and Dominica, respectively. Other species, mostly from the larger islands, have been recorded more widely.

Populations of the white-necked crow *(Corvus leucognaphalus - VU)*, golden swallow (VU), Hispaniolan parakeet (*Amazona ventralis* - VU) and Hispanio an parakeet *(Psittacara chloropterus* - VU) have been confirmed at more than 10 sites in the Greater Antilles.

Table 5.4 Globally threatened birds by country and number of CBAs.

Tabla 5.4 Aves amenazadas a nivel mundial por país y número de ACB (solo países elegibles del CEPF)

Familia	Especie	Lista Roja de la UICN	Antigua y Barbuda	Bahamas	Dominica	República Dominicana	Granada	Haití	Jamaica	Santa Lucía	San Vicente y las Granadinas
Accipitridae	Halcón de Ridgway (*Buteo ridgwayi*)	CR				2					
Anatidae	Pato silbador de las Indias Occidentales (*Dendrocygna arborea*)	VU	5	1					3		
Calyptophilidae	Angara haitiana/ Chirrí de Bahoruco (*Calyptophilus tertius*)	VU				2		4			
Columbidae	Paloma perdiz dominicana/perdiz oculta (*Geotrygon leucometopia*)	EN				2					
	Paloma montaraz de Granada (*Leptotila wellsi*)	CR					5				
	Paloma jamaicana (*Patagioenas caribaea*)	VU							7		
Corvidae	El cuervo de la Española/cuervo de cuello blanco (*Corvus leucognaphalus*)	VU				10		6			
Cuculidae	Cuco picogordo de la Española (*Coccyzus rufigularis*)	EN				5					
Fringillidae	El piquituerto de la Española (*Loxia megaplaga*)	EN				4		4			
Hirundinidae	Golondrina de Las Bahamas (*Tachycineta cyaneoviridis*)	EN		2							
	Golondrina dorada (*Tachycineta euchrysea*)	VU				7		4			
Icteridae	Turpial de Las Bahamas (*Icterus northropi*)	CR		5							
	Zanate jamaicano (*Nesopsar nigerrimus*)	EN							4		
Mimidae	Cuitlacoche pechiblanco (*Ramphocinclus brachyurus*)	EN								2	
Parulidae	Reinita de San Vicente (*Catharopeza bishopi*)	EN									7
Phaenicophilidae	Reinita montana / Cigüita albianca (*Xenoligea montana*)	VU				7		4			
Procellariidae	Petrel diablotín (*Pterodroma hasitata*)	EN				1		3			
Psittacidae	Amazona de pico negro (*Amazona agilis*)	VU							5		
	Amazona gorgirroja (*Amazona arausiaca*)	VU			2						
	Amazona de Pico Amarillo (*Amazona collaria*)	VU							8		
	Amazona de San Vicente (*Amazona guildingii*)	VU									7

3.1.3 Reptiles

Of the 186 threatened reptile species in the Caribbean, 118 occur in countries eligible for CEPF investment. Of these, 57 have been recorded in at least one CBA. Fifty-two reptiles trigger CBA criteria: 21 Critically Endangered; 18 Endangered; and 13 Vulnerable (Table 5.5).

Several reptile species have not been proposed for any CBA, it should be recognized that this is an information gap in the ecosystem profile, given the large number of endemic and threatened reptile species. This gap can be partially explained by two things. First, although a major Red List assessment was published for this group in 2017, the results may not yet have been incorporated into practice at the time of ecosystem profile consultations, resulting in some species now classified as threatened having gone unnoticed and not being proposed as CBA-triggering species.

Secondly, there is relatively little information available for some species compared to other groups, although this is changing rapidly, particularly in several countries of the Lesser Antilles, such as Antigua and Barbuda, Barbados and St. Lucia, where there are important efforts to conserve these species.

Families underrepresented in this update include Amphisbaenidae, Anguidae, Dipasadidae, Leptothyphlopidae, Scinidae, Sphaerodactylidae (the family with the highest number of missing species), Tropiduridae and Typhlopidae.

Groups adequately represented in the CBAs include better known families, such as turtles and iguanas, which had been assessed by IUCN previously and were therefore also included in the previous ecosystem profile.

As expected due to the nature of some species (many of them restricted to individual islands), most reptiles in the Caribbean have been reported from one or very few sites, with the exception of turtles (mostly marine but also one terrestrial species), some boas and snakes, many of the iguanas, the American crocodile *(Crocodylus acutus) and the Trachemys* tortoises. A total of 11 Critically Endangered or Endangered reptile species are confined to a single site and therefore trigger AZE sites: Cayemite Long-tailed Amphisbaena - *(Amphisbaena caudalis* - EN), Cayemite Short-tailed Amphisbaena (*A. cayemite* - CR), the silver boa of the Concepción bank *(Chilabothrus argentum* - CR), the Redonda lizard *(Anolis nubilis* - CR), the Antigua snake *(Alsophis antiguae* - CR), the St. Lucia snake *(Erythrolamprus ornatus* - CR), the Jamaican iguana *(Cyclura collei* - CR), the Union Island gecko *(Gonatodes daudini* - CR), the spotted agave spherodactyl *(Sphaerodactylus ladae* - EN), the Redonda ameiva *(Pholidoscelis atratus* - CR) and the High Veil leiocephalus *(Leiocephalus altavelensis* - CR).

Table 5.5 Globally threatened reptiles by country and number of CBAs.

Family	Species	IUCN Red List
Amphisbaenidae	Cayemite Long-tailed Amphisbaena (*Amphisbaena caudalis*)	EN
	Cayemite Short-tailed Amphisbaena (*Amphisbaena cayemite*)	CR
Anguidae	Fowler's celestus (*Celestus fowleri*)	VU
	Hispaniolan Giant Lizard (*Celestus warreni*)	VU
Boidae	Concepción bank silver boa (*Chilabothrus argentum*)	CR
	Jamaican Boa (*Chilabothrus subflavus*)	VU
Cheloniidae	Loggerhead sea turtle (*Caretta caretta*)	VU
	Hawksbill turtle (*Eretmochelys imbricata*)	CR
Colubridae	St Vincent Blacksnake (*Chironius vincenti*)	CR
Crocodylidae	American crocodile (*Crocodylus acutus*)	VU
Dactyloidae	Stout Anole Shark (*Anolis haetianus*)	EN
	St. Lucia's Anole (*Anolis luciae*)	EN
	Round-tailed Lizard (*Anolis nubilis*)	CR
Dermochelyidae	Leatherback Turtle (*Dermochelys coriacea*)	VU
Dipsadidae	Antigua snake (*Alsophis antiguae*)	CR
	St. Lucia Snake (*Erythrolamprus ornatus*)	CR
	Brown racer snake (*Haitiophis anomalus*)	VU
	(*Laltris agyrtes*)	EN
Emydidae	Spanish Sea Turtle (*Trachemys decorata*)	VU
	Cat Island Freshwater Turtle (*Trachemys terrapen*)	VU
Iguanidae	Turks and Caicos Iguana (*Cyclura carinata*)	CR
	Jamaican Iguana (*Cyclura collei*)	CR
	Rhinoceros iguana (*Cyclura cornuta*)	VU

Family	Species	IUCN Red List
	Northern Bahama Iguana *(Cyclura cychlura)*	VU
	Ricord's Iguana *(Cyclura ricordii)*	CR
	Bahamas Iguana *(Cyclura rileyi)*	EN
	Caribbean Iguana *(Iguana delicatissima)*	EN
Leiocephalidae	East Plana Curlytail Lizard *(Leiocephalus greenwayi)*	VU
Leptotyphlopidae	Martin Garcia's Garter Snake *(Mitophis asbolepis)*	CR
	Samana Garter Snake *(Mitophis calypso)*	CR
	St. Lucia Garter Snake *(Tetracheilostoma breuili)*	EN
	Barbados Garter Snake *(Tetracheilostoma carlae)*	CR
Phyllodactylidae	Dominican Leaf-toed Gecko *(Phyllodactylus hispaniolae)*	EN
	Barbados Leaf-toed Gecko *(Phyllodactylus pulcher)*	CR
Scincidae	Jamaican Skink *(Spondylurus fulgida)*	EN
Sphaerodactylidae	Union Island Gecko *(Gonatodes daudini)*	CR
	Striped spherodactyl of the Hyatids *(Sphaerodactylus cochranae)*	CR
	Bakoruco Least Gecko *(Sphaerodactylus cryphius)*	EN
	Grenadines Sphaero *(Sphaerodactylus kirbyi)*	VU
	Spotted agave spherodactyl *(Sphaerodactylus ladae)*	EN
	Martin Garcia's Spherodactylus *(Sphaerodactylus perissodactylius)*	EN
	Pedernales Least Gecko *(Sphaerodactylus randi)*	EN
	Haitian banded spherodactyl *(Sphaerodactylus samanensis)*	CR
	Neiba agave spherodactyl *(Sphaerodactylus schuberti)*	CR
	Sphaerodactylus semasiops	EN

Family	Species	IUCN Red List
	Spherodactyl of the Barahona limestones *(Sphaerodactylus thompsoni)*	EN
Teiidae	St. Lucia whip lizard *(Cnemidophorus vanzoi)*	CR
	Ameiva of Redonda *(Pholidoscelis atratus)*	CR
Tropiduridae	High-veiled Leiocephalus *(Leiocephalus altavelensis)*	CR
Typhlopidae	Grenada Bank Blindsnake *(Amerotyphlops tasymicris)*	EN
	Barahona Peninsula Blind Snake *(Typhlops syntherus)*	EN
Viperidae	Santa Lucia spearhead *(Bothrops caribbaeus)*	EN

Seventy-eight of the 146 threatened Caribbean amphibian species are found in eligible countries. Sixty-one of these species trigger CBA criteria.

The threat status of these species is as follows: 29 Critically Endangered, 24 Endangered and eight Vulnerable (Table 5.6). As with reptiles, there are notable information gaps for amphibians in this profile: the presence of 19 threatened amphibian species has not been confirmed in any CBA.

Amphibian species not confirmed in any CBA include two members of the family Bufonidade (endemic to the Dominican Republic), three members of Dactyloidae (endemic to The Bahamas, St. Lucia, and Antigua and Barbuda), and 14 species of Eleutherodactylidae, endemic to Hispaniola (nine only to Haiti) and Jamaica.

Amphibians have highly restricted ranges in the hotspot, so most species are restricted to fewer than three CBAs, with some notable exceptions in Jamaica and Hispaniola, where some species have been recorded from as many as 20 sites. Given the relatively large size of the Greater Antilles, this situation is to be expected.

Two Critically Endangered amphibian species are restricted to a single site and therefore trigger AZE criteria: *Eleutherodactylus caribe* and *E. sisyphodemus.*

Table 5.6 Globally threatened amphibians by country and number of CBAs (CEPF-eligible countries only)

Family	Species	IUCN Red List
Bufonidae	Southern Crested Toad *(Peltophryne guentheri)*	VU
Craugastoridae	*Pristimantis euphronides*	EN
Pristimantis shrevei		EN

Family	Species	Red List of the IUCN
Eleutherodactylidae	Barahona Rock Frog *(Eleutherodactylus alcoae)*	EN
Mozart's Frog *(Eleutherodactylus amadeus)*		CR
Eleutherodactylus amplinympha		EN
Jamaican Rumpspot Frog *(Eleutherodactylus andrewsi)*		EN
La Hotte's big-legged frog *(Eleutherodactylus apostates)*		CR
Baoruco hammerhead frog *(Eleutherodactylus armstrongi)*		EN
Hispaniolan telegraph frog *(Eleutherodactylus audanti)*		VU
Northern hammerhead frog *(Eleutherodactylus auriculatoides)*		EN
Green flat frog *(Eleutherodactylus brevirostris)*		CR
Haitian Marsh Frog *(Eleutherodactylus caribe)*		CR
Eleutherodactylus cave dweller		CR
Eleutherodactylus corona		CR
Eleutherodactylus counouspeus		EN

Eleutherodactylus dolomedes	CR
Eleutherodactylus eunaster	CR
Bromeliad kakhi frog *(Eleutherodactylus fowleri)*	CR
Red-legged frog of La Selle *(Eleutherodactylus furcyensis)*	CR
Eleutherodactylus fuscus	CR
La Hotte's Gland Frog *(Eleutherodactylus glandulifer)*	CR
Eleutherodactylus glaphycompus	EN
Eleutherodactylus grabhami	EN
Eleutherodactylus griphus	CR
Mountain cricket frog *(Eleutherodactylus haitianus)*	EN
Bromeliad midline frog *(Eleutherodactylus heminota)*	EN
Bahoruco burrowing frog *(Eleutherodactylus hypostenor)*	EN
Eleutherodactylus jamaicensis	EN
La Selle Brown Frog *(Eleutherodactylus jugans)*	CR
Eleutherodactylus junori	CR
Southern pie frog *(Eleutherodactylus leoncei)*	CR
Eleutherodactylus luteolus	EN
Hispaniolan Sobting Frog *(Eleutherodactylus minutus)*	EN

Family	Species	IUCN Red List
Hispaniolan Mountain Frog *(Eleutherodactylus montanus)*		EN
Green spiny frog *(Eleutherodactylus nortoni)*		CR
Eleutherodactylus oxyrhyncus		CR
Neiba whistling frog *(Eleutherodactylus parabates)*		CR
Eleutherodactylus parapelates		CR
Hispaniolan Mountain Frog *(Eleutherodactylus patriciae)*		EN
Eleutherodactylus paulsoni		CR
Eleutherodactylus pentasyringos		VU
Hispaniolan Yellow Spotted Frog *(Eleutherodactylus pictissimus)*		VU
Hispaniolan Melodious Frog *(Eleutherodactylus pituinus)*		EN
Eleutherodactylus poolei		CR
Eleutherodactylus probolaeus		EN
Eleutherodactylus rhodesi		CR
Bahoruco Red-legged Frog *(Eleutherodactylus rufifemoralis)*		CR
Eastern Burrowing Frog *(Eleutherodactylus ruthae)*		EN
Eleutherodactylus semipalmatus		CR
Eleutherodactylus sisyphodemus		CR
Eleutherodactylus thorectes		CR
Macayan dusky frog Eleutherodactylus ventrilineatus		CR
Haitian Whistler Frog *(Eleutherodactylus wetmorei)*		VU
Hylidae Hispaniolan green tree frog *(Hypsiboas heilprini)*		VU
Crucial Osteopilus		EN
Osteopilus marianae		EN
Hispaniolan Yellow Tree Frog *(Osteopilus pulchrilineatus)*		VU

Hispaniolan Giant Tree Frog *(Osteopilus vastus)*		VU
Osteopilus wilderi		EN
Leptodactylidae	Mountain chicken *(Leptodactylus fallax)*	CR

3.1.5 Freshwater and coastal marine fish

There is still an important gap in the knowledge of the biodiversity of bony fishes in the Caribbean. There are about 1 600 species in the region. Twenty-nine of the 33 threatened fish species occurring in the eligible countries are not endemic to the Caribbean and have wide distribution and some commercial value (tuna, seahorses and groupers, etc.).

Many endemic fish species have not been assessed under the Red List criteria (many species of Poeciliidae) or the available assessments need to be updated (e.g., mosquitofish or gambusia was last assessed in 2009 and Nassau grouper in 2003).

The application of the CBA criteria in this profile was limited to globally threatened species, but only approximately 2 % of all fish species globally are classified as threatened. The current results for fish species should therefore be considered preliminary.

Of the 33 globally threatened fish species in CEPF eligible countries, only five species (four Endangered and one Vulnerable) trigger CBA criteria (Table 5.7).

Table 5.7 Globally threatened fishes by country and number of CBAs (CEPF eligible countries only).

Family	Species	IUCN Red List
Anguillidae	American Eel *(Anguilla rostrata)*	EN
Bythitidae	*Lucifuga lucayana*	EN
Bythitidae	Bahamian loosestrife *(Lucifuga spelaeotes)*	VU
Epinephelidae	Striated grouper *(Epinephelus striatus)*	EN
Poeciliidae	Mosquito fish or gambusia *(Gambusia dominicensis)*	EN

3.1.6 Cartilaginous fish

Out of a total of 16 threatened cartilaginous fish species found in CEPF eligible countries, only one species, the small sawfish (CR), triggered a CBA during this profiling exercise. As with bony fishes, there are significant gaps in knowledge of cartilaginous fishes in hotspot CBAs.

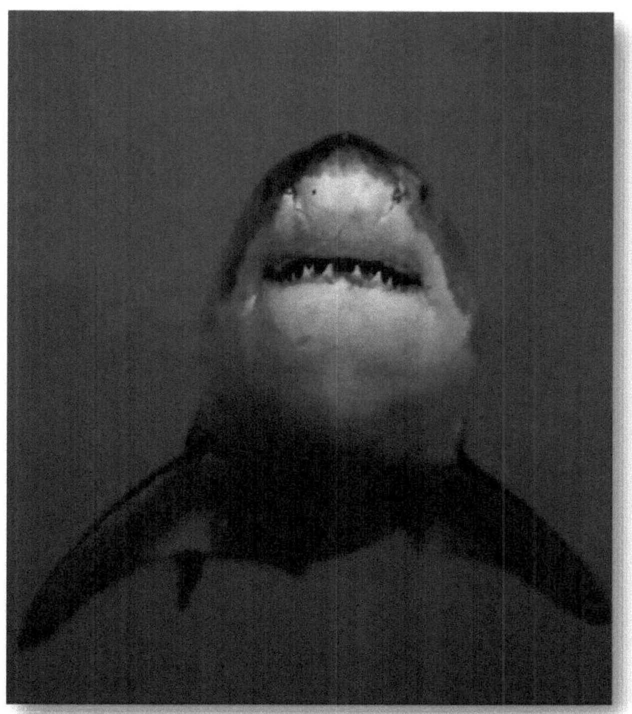

3.1.7 Reef-building corals

There are 11 threatened species of corals (Anthozoa) and fire corals (Hydrozoa) in CEPF eligible countries. Nine have been reported in the CBAs in Antigua and Barbuda, The Bahamas, Dominican Republic and Haiti: two Critically Endangered, one Endangered and six Vulnerable.

Despite the importance of corals in the region, several data-related problems prevented the application of CBA criteria during the ecosystem profile preparation process. Therefore, no CBA was triggered by any coral species.

Some of the sites proposed by stakeholders as CBAs for coral species, such as the North East Marine Management Area and

Fitches Creek Bay sites in Antigua and Barbuda, Aire Protegée de Ressources Naturelles Gérées des Trois Baies in Haiti, and Exuma Cays Land and Sea Park in The Bahamas will likely trigger coral-based CBA criteria once data issues are resolved. Other coastal sites not triggered as CBAs by any other species at this time will likely be added to the CBA landscape in the near future.

3.2 Fauna by site

To date, a total of 324 CBAs have been identified in the West Indies biodiversity hotspot, 167 of which are in CEPF eligible countries (Table 5.9 and Figure 5.1). These sites were identified at different times using different methodologies. As a result, there are currently four different datasets for Caribbean CBAs: (i) CEPF eligible countries; (ii) EU OCTs; (iii) Cuba; and (iv) Puerto Rico and the U.S. Virgin Islands. Sites in Cuba, EU OCTs, and Puerto Rico and the U.S. Virgin Islands were identified prior to the introduction of the new CBA Standard (IUCN 2016).

At some point in the future, these CBAs should be reassessed against the new CBA Standard to resolve their global/regional status.

Eligible CEPF countries. One hundred and sixty-seven CBAs were identified in the 11 CEPF eligible countries (Figures 5.2 to 5.7). The vast majority (157) of these sites were identified as CBAs in CEPF's previous ecosystem profiling process (CEPF, 2010).

The implementation of the new CBA Standard is a multi-step process, involving pre-assessment, expert review and confirmation by the CBA Secretariat.

It was not possible to complete all of these steps during the ecosystem profile update process. Therefore, although all of these sites qualify as CBAs, the global/regional status of each remains to be confirmed. Final confirmation of the status of these CBAs will only occur when they are entered into the global CBA database (http://www.keybiodiversityareas.org); additional expert review may be required at this time.

Table 5.9 Summary of key faunal areasq by country of the Antilles biodiversity hotspot.

CEPF Eligible Countries Country/Quantity	CBA 2009	CBA 2018
Antigua and Barbuda	10,0	6,0
Bahamas	26,0	23,0
Barbados	4,0	7,0
Dominica	4,0	4,0
Dominican Republic	35,0	39,0
Grenada	9,0	9,0
Haiti	17,0	30,0
Jamaica	38,0	32,0
St. Kitts and Nev s	1,0	2,0
St. Lucia	6,0	7,0
Saint Vincent and the Grenadines	7,0	8,0
Subtotal for CEPF eligible countries:	**157,0**	**167,0**
Cuba	28,0	28,0
U.S. overseas territories		
Puerto Rico	28,0	27,0
U.S. Virgin Islands	13,0	11,0
EU OCT	**France**	
Guadalupe	8,0	10,0
Martinique	8,0	8,0
St. Bartholomew's	4,0	3,0
St. Maarten	1,0	2,0
Netherlands		
Aruba	1,0	7,0
Bonaire	4,0	6,0

Curaçao	0,0	6,0
Saba	1,0	4,0
St. Eustace	2,0	3,0
Sint Maarten	0,0	5,0
United Kingdom		
Eel	6,0	5,0
Cayman Islands	8,0	8,0
Montserrat	3,0	6,0
Turks and Caicos Islands	11,0	11,0
Virgin Islands	7,0	7,0
Total	**290,0**	**324,0**

Figure 5.1. Key biodiversity areas in the West Indies biodiversity hotspot.

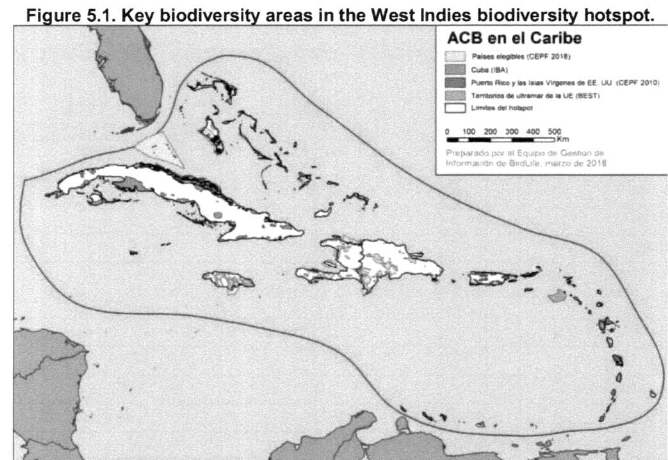

The analysis for this CEPF ecosystem profile update follows the recently adopted CBA Standard (IUCN 2016). As recommended by the CBA Standard, the baseline for the CBA list takes into account new proposals as well as existing initiative sites, such as:
• CBAs defined using earlier criteria (e.g., Langhammer *et al.* 2007), such as those defined in the previous phase of CEPF's investment in the West Indies (CEPF 2010).
• IBA and AZE sites.
• Protected areas.

After a desktop analysis, a preliminary list of sites to be reviewed as CBAs was generated and shared with national experts (electronically and through an interactive ArcGIS Story Map microsite) and discussed during three national workshops (Dominican Republic, Haiti, and Jamaica) and through an online sub-regional consultation for The Bahamas and the Eastern Caribbean (Brown *et al.*, 2019).

During this process, national stakeholders reviewed existing information and provided new data, including site polygons, species records and references.

Following the previous evaluation of sites as CBAs (see Section 5.2.1), the list of CBAs with the highest biological values was subsequently reviewed by the national expert groups and the participants of the regional workshop in Jamaica (Brown *et al., ob. cit.).*

EU OCTs. Ninety-two sites were identified in EU OCTs through the BEST initiative and are documented in a specific ecosystem profile (Vaslet and Renoux 2016, see Figures 5.8 to 5.10).

The previous CEPF ecosystem profile identified 64 CBAs in these countries. The 92 sites identified through the BEST initiative followed the above CBA criteria (Langhammer *et al.,* 2007).

Due to differences in criteria and methodology between the previous CBA criteria and the current CBA standard, the results of the BEST process cannot be directly compared to the dataset analyzed in this paper for CEPF eligible Caribbean countries. Details of the OCT sites can be found at: http://ec.europa.eu/environment/nature/biodiversity/best/regions/cari bbean_e n.htm.

Cuba. There are 28 CBAs in Cuba (Figure 5.11), all based on IBAs identified in 2008, so there are no changes to the information presented in the previous ecosystem profile (2010). This CBA dataset is limited to birds only because it was not possible to conduct a detailed CBA analysis for Cuba that incorporates other taxonomic groups.

It is important to note that IBAs qualify as CBAs because of their importance to global biodiversity and that the sites provide habitat for other species that may also have global biodiversity conservation significance. Due to differences in criteria and methodology, and the absence of details on taxa other than birds, a comparative analysis cannot be made with the CBAs included in this document for CEPF eligible countries.

Details of these 28 Cuban CBAs and their trigger species can be found in the World Database of Key Biodiversity Areas: http://www.keybiodiversityareas.org/site/results?reg=4andcty=53and snm.

Puerto Rico and the U.S. Virgin Islands. Although this profile update focused on CEPF eligible countries, minor updates have been made for Puerto Rico and the U.S. Virgin Islands (Figure 5.12).
Puerto Rico currently has 27 CBAs (down from 28 in 2010), six sites do not qualify as CBAs and another six sites lack the necessary information to be evaluated as CBAs.
There are 11 CBAs in the U.S. Virgin Islands, while eight others require more information to be evaluated.

The number of CBAs in the hotspot could change as soon as additional information on sites and species not included in this profile becomes available, e.g., once Criteria B is applied or when some of the knowledge gaps identified in this analysis are filled (e.g., reptiles, some amphibian families, corals, fishes, and other taxonomic groups not reviewed in this profile). Therefore, it is expected that some additional sites will be added to the list of CBAs in the hotspot in the future.
CBAs range in size from the small Bethesda Dam area in Antigua and Barbuda (less than 2 hectares) to the huge Cay Sal Marine Management Area in The Bahamas (over 1.6 million hectares).
The average size for CBAs is 4 500 hectares, but the average size at the national level varies from country to country. Among CEPF-eligible countries, CBAs cover from 3.6% of the total area (St. Kitts and Nevis) to 37.1% (St. Lucia).
The marine area covered by CBAs is much smaller, with an average coverage of 1.6 % of the Exclusive Economic Zone (EEZ) of CEPF eligible countries (Table 5.10).
The countries with the highest percentage of land area covered by CBA are St. Lucia, Jamaica, St. Vincent and the Grenadines, Dominica and The Bahamas.
In terms of area, The Bahamas stands out above the Dominican Republic, Haiti and Jamaica, thanks to the Cay Sal Marine

Management Area. As expected for an island country, the marine area of The Bahamas is notably larger than the terrestrial area (its EEZ covers 1.7 million km2 compared to 104 000 km^2 of land area). However, the figures for the area covered by CBAs in terrestrial and marine environments are relatively similar (approximately 27 000 km^2 of marine area versus 21 000 km^2 of terrestrial area.

Table 5.10 Land and marine area of CBAs by country (CEPF eligible countries only).

Country	Land area of the country (km2)	Marine area (km2)	Total area of CBA (km2)	CBA land area (km2)	CBA marine area (km2)	CBA land coverage (%)	CBA marine coverage (%)
Antigua and Barbuda	440,0	111 914	202,0	58,0	144,0	13,2	0,1
Bahamas	13 880	619 938	24 154	3 988	20 166	28,7	3,3
Barbados	430,0	185 704	68,0	67,0	1,0	15,6	0,0
Dominica	750,0	28 653	229,0	224,0	5,0	29,9	0,0
Dominican Republic	48 730	351 756	9 576	8 198	1 378	16,8	0,4
Grenada	340,0	25 670	33,0	33,0	0,0	9,7	0,0
Haiti	27 750	103 818	8 550	4 749	3 802	17,1	3,7
Jamaica	10 990	257 777	5 546	3 900	1 646	35,5	0,6
St. Kitts and Nevis	360,0	9 533	13,0	13,0	0,0	3,6	0,0
St. Lucia	620,0	15 470	247,0	230,0	17,0	37,1	0,1
Saint Vincent and the Grenadines	390,0	36 381	135,0	134,0	1,0	34,4	0,0
Total	104 680	1 746 614	48 753	21 594	27 160	20,6	1,6

About 20% of the land area of all countries is under some form of formal protection, the figures for marine protected area are lower (about 6% of the total EEZ area). Only 1% of Barbados is under some form of formal protection, while at the other end of the spectrum, The Bahamas has 35% of its land area covered by protected areas, followed by the Dominican Republic with 25%.

The protection status of CBAs in CEPF-eligible countries is relatively high (Table 5.11). Seventy-nine percent of these CBAs overlap with some type of protected area. This is to be expected, given the considerable amount of biological information for threatened species available for these sites. However, there are differences in the status of protection as CBAs among CEPF-eligible countries, some having higher levels of protection and others lower.

Eighty percent of the CBAs in the Dominican Republic, The Bahamas and Antigua and Barbuda are under some form of legal protection, while in St. Kitts and Nevis and Barbados, less than 2 percent of the CBAs are covered by protected areas.

In Haiti, 26% of CBAs enjoy some form of formal protection. In the latter three countries, there are opportunities to use CBAs to support the identification of protected areas or 'Other Effective Area-based Conservation Measures' (OECM).

Among CEPF-eligible countries, The Bahamas, Dominican Republic, St. Vincent and the Grenadines, Dominica, and Antigua and Barbuda have met Aichi Target No. 11 in inland and terrestrial waters (17% of territory under some form of protection, including OECM). Only the Dominican Republic meets the coastal marine target (10%), which positions CBAs as a useful tool to support the achievement of this target in other hotspot countries.

More than 100 protected areas in the hotspot have not yet been confirmed as CBAs due to lack of information. It is important to assess protected areas in future CBA assessments, but it is equally important to assess and identify CBAs outside of national protected area systems.

Table 5.11 Surface area of protected area and CBA (terrestrial and marine) (CEPF eligible countries only).

Country	Total protected area (ha)	Protected land area (ha)	Marine protected area (ha)	Total CBA area (ha)	Total area of CBA under protection (ha)	Land area of CBA under protection (ha)	Marine area CBA under protection (ha)
Antigua and Barbuda	25 553	8 118	17 435	20 200	17 594	5 291	12 303
Bahamas	5 208 792	487 051	4 721 740	2 415 400	2 301 403	321 294	1 980 109
Barbados	1 559	464,0	1 095	6 800	122,0	109,0	13,0
Dominica	17 196	16 139	1 058	22 900	10 066	10 065	0,0
Republic Dominican	6 059 728	1 216 417	4 843 310	957 600	943 690	807 079	136 611
Grenada	4 043	3 038	1 005	3 300	2 176	2 176	0,0
Haiti	372 870	187 214	185 656	855 000	225 457	88 847	136 611
Jamaica	361 004	175 495	185 509	554 600	321 732	160 342	161 390
San Cristóbal and Nieves	n/a	n/a	0n/a	1 300	n/a	n/a	n/a
St. Lucia	13 431	9 758	3 673	24 700	10 879	9 643	1 236

St. Vincent and the Grenadines	17 187	8 928	8 259	13 500	7 387	7 387	0,0
Total	12 081 363	2 112 622	9 968 740	4 875 300	3 840 506	1 412 234	2 428 272

Of the 167 CBAs in CEPF-eligible countries, 93 sites are reptile-activated (Table 5.12). Seventy-three sites are activated by birds, 55 by amphibians, 46 by mammals, seven by fish (both freshwater and marine) and one by sharks. Corals did not activate any CBAs (see section 5.1.7).

Table 5.12 Summary of key biodiversity areas by taxonomic group in CEPF eligible countries.

Taxonomic group	Number of CBAs activated	Percentage from ACB activated
Mammals	46,0	28,0
Birds	73,0	44,0
Reptiles	93,0	56,0
Amphibians	55,0	33,0
Fish	7,0	4,0
Sharks	1,0	<1,0
Plants with seed	85,0	51,0
All CBAs	167,0*	N/A

On average, each site is activated by five or more species. However, some CBAs support exceptional numbers of globally threatened species including Cockpit Country, Blue and John Crow Mountains Protected National Heritage area and its environs, and Litchfield Mountain-Matheson's Run in Jamaica, and Sierra de Bahoruco National Park in the Dominican Republic. Each of these sites has more than 30 trigger species.

Seventeen CBAs are considered totally irreplaceable on a global scale because they contain the only known population of one or more globally threatened species (Table 5.13). Because all of these sites are irreplaceable for Critically Endangered and Endangered species they also qualify as AZE sites: the most urgent conservation priorities globally. It is important to note an important difference between the

AZE and CBA criteria: Critically Endangered (possibly extinct) species can trigger the identification of an AZE site, but such species cannot be used for the identification of a CBA. This, plus some discrepancies in site delimitation, explains the differences between the updated AZE sites and the current dataset of irreplaceable CBAs in eligible countries.

At the time of completion of this paper, the authors (Brown *et al., ob. cit.),* were aware of at least two confirmed AZE sites not included in this profile. A small area of Île la Tortue, Haiti, is reported to be home to a population of Warren's robber frog *(Eleutherodactylus warren -* CR), but this species has not been re-reported there since it was first described; the above may preclude confirmation of this site as an ACB. For this reason, the site is not listed as a CBA in this profile.

Table 5.13 Totally irreplaceable sites in the West Indies hotspot (CEPF-eligible countries only)

Country	Site name	Species
Antigua and Barbuda	North East Marine Management Area and Fitches Creek Bay	Antigua snake *(Alsophis antiguae)*
Round	Redondo's Lizard *(Anolis nubilis);* Redonda's Ameiva *(Pholidoscelis atratus)*	
Bahamas	Conception Island National Park	Bank silver boa Concepción *(Chilabothrus argentum)*
Dominican Republic	Jaragua National Park	High Veil Leiocephalus *(Leiocephalus altavelensis)*
Sierra Martin National Park Garcia	Spotted agave spherodactylus *(Sphaerodactylus ladae)*	
Haiti	Cayemites - Barradères	Cayemite Long-tailed Amphisbaena *(Amphisbaena caudalis);* Cayemite Short-tailed Amphisbaena *(A. cayemite)*
Dame Marie	*Caribbean Coqui (Eleutherodactylus Caribe)*	
Jamaica	Cockpit Country Jamaican Coqui *(Eleutherodactylus sisyphodemus)*	
	Portland Bight Protected Area Jamaican Iguana *(Cyclura collei)*	
St. Lucia	Pointe Sable	*Erythrolamprus ornatus*
Saint Vincent and the Grenadines	Chatham Bay, Union Island	*Gonatodes daudini*

Figures 5.2 to 5.7 show the location of site results (CBAs) in each of

the CEPF eligible countries. Because the completeness of available data on the distribution of globally threatened species among CBAs varies significantly among taxonomic groups, CBAs identified as important for the conservation of one taxonomic group may also be important for other groups for which data are not yet available or for which data were insufficient at the time of the previous CBA assessment. In addition, there are likely to be other globally threatened species conservation important sites in the region that were not identified during the ecosystem profile update process, especially for plants, reptiles, and marine species, as well as for non-threatened species that may trigger other CBA criteria, such as non-threatened endemic species and congregating species. It is hoped that additional analysis may fill these data gaps in the future.

3.2.2 Site results for EU Overseas Countries and Territories and Outermost Regions

The Caribbean region contains several overseas countries and territories and outermost regions of the EU member states, France, the Netherlands and the United Kingdom. As more than 70% of Europe's species are found in EU overseas countries and territories and outermost regions, the biodiversity of these sites has been recognized as being of international importance and crucial to achieving global and regional biodiversity targets.

There are 15 EU islands and island groups in the Antilles biodiversity hotspot. These are the Dutch islands of Aruba, Bonaire, Curaçao, Saba, St. Eustatius and Sint Maarten; the French islands of Guadeloupe, Martinique, St. Martin and St. Barthelemy; and the British islands of Anguilla, British Virgin Islands, Cayman Islands, Montserrat and the Turks and Caicos Islands.

Like the rest of the hotspot, these islands have very diverse ecosystems and biomes, as a result of diverse climatic, topographic, geological and biogeographic patterns (Petit and Prudent, 2010). They include wetlands (mangroves, etc.), seagrass beds, coral reefs, beaches, rivers and streams, tropical grasslands, savannas and scrublands, tropical dry forests and rainforests.

The regional ecosystem profile prepared under the BEST initiative identified 92 CBAs, including 42 terrestrial CBAs and 50 marine and coastal CBAs (Vaslet and Renoux, 2016). These include 31 CBAs in the Dutch OCTs, 24 in the French Outermost Regions (ORs) and Overseas Territories (UTs), and 37 in the UK UTs, covering a combined area of 8 090 km^2 . These CBAs take into account 194 globally threatened species, 1094 endemic and restricted-range species and around 45 species that congregate in significant numbers to feed or breed, mainly represented by birds and marine mammals. The species list includes 173 vertebrate species, more than 430 invertebrate species and 488 plant species.

EU BEST CBAs cover freshwater, coastal, marine and terrestrial ecosystems. The EU overseas entities of the West Indies hotspot host two sites identified in the 2010 AZE assessment.

Montserrat's Centre Hills area was designated as an AZE site due to the presence of the endemic and threatened Montserrat oriole *(Icterus oberi), while* the forest ecosystem in Basse-Terre (Guadeloupe) was designated due to the presence of endemic and threatened amphibian species.

Figure 5.2. Site results in The Bahamas

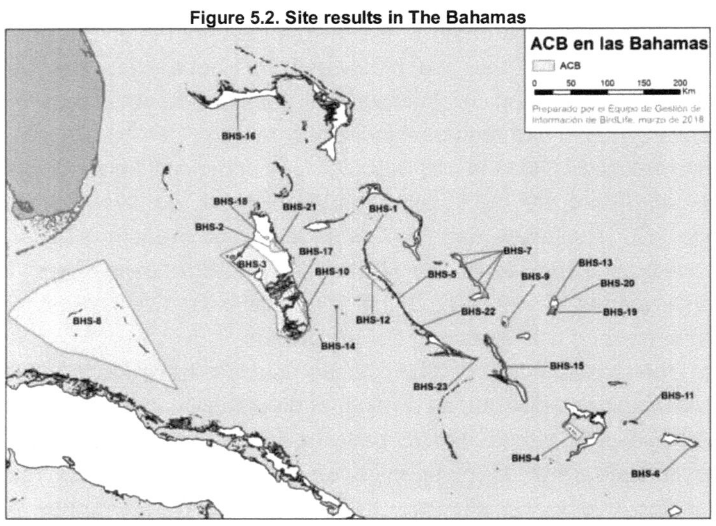

Figure 5.3. Site results in Jamaica

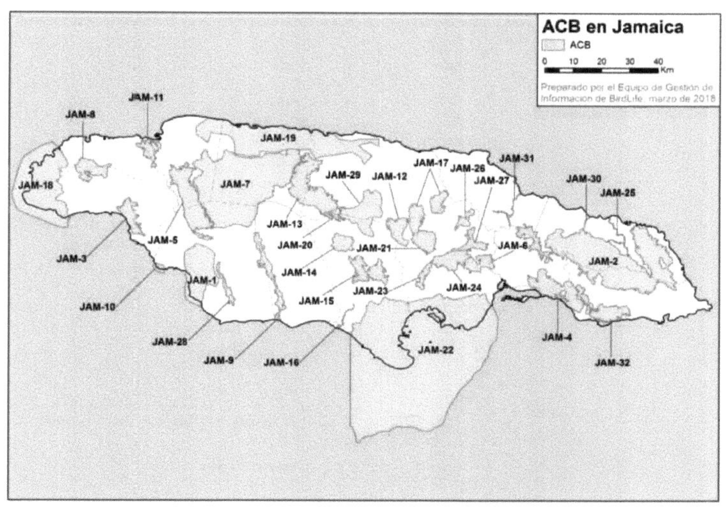

Figure 5.4. Haiti site results

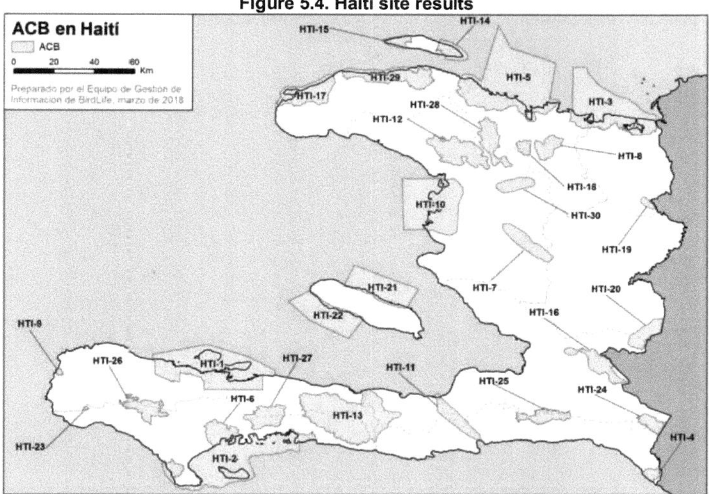

Figure 5.5. Site results in Dominican Republic

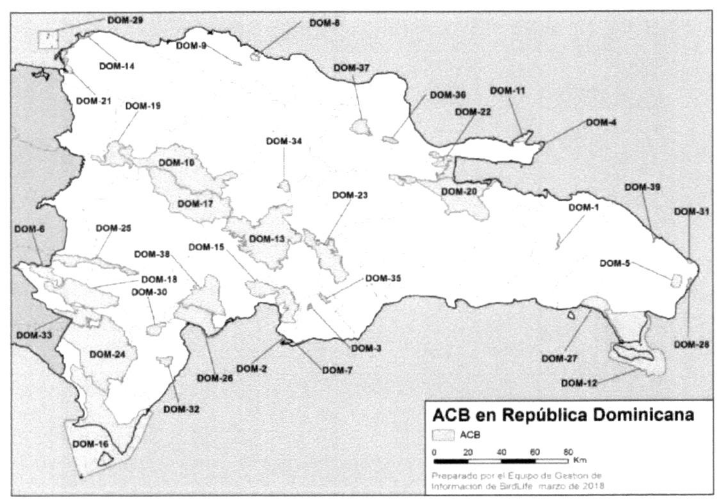

Figure 5.6. Site results in St. Kitts and Nevis, Antigua and Barbuda, and Dominica.

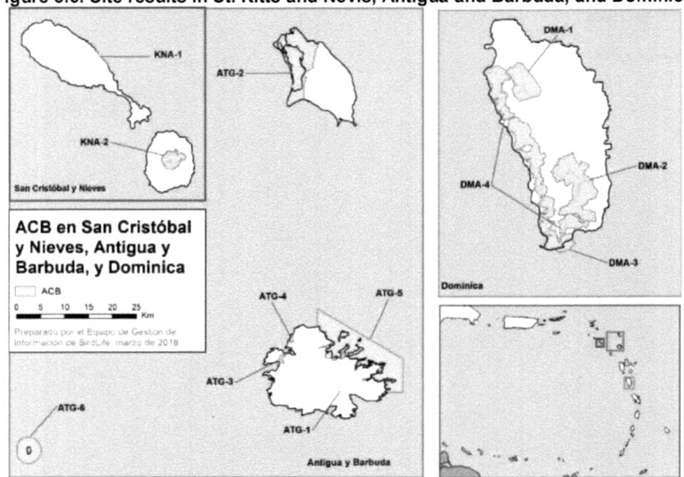

Figure 5.7. Site results in St. Lucia, Barbados, St. Vincent and the Grenadines, and Grenada.

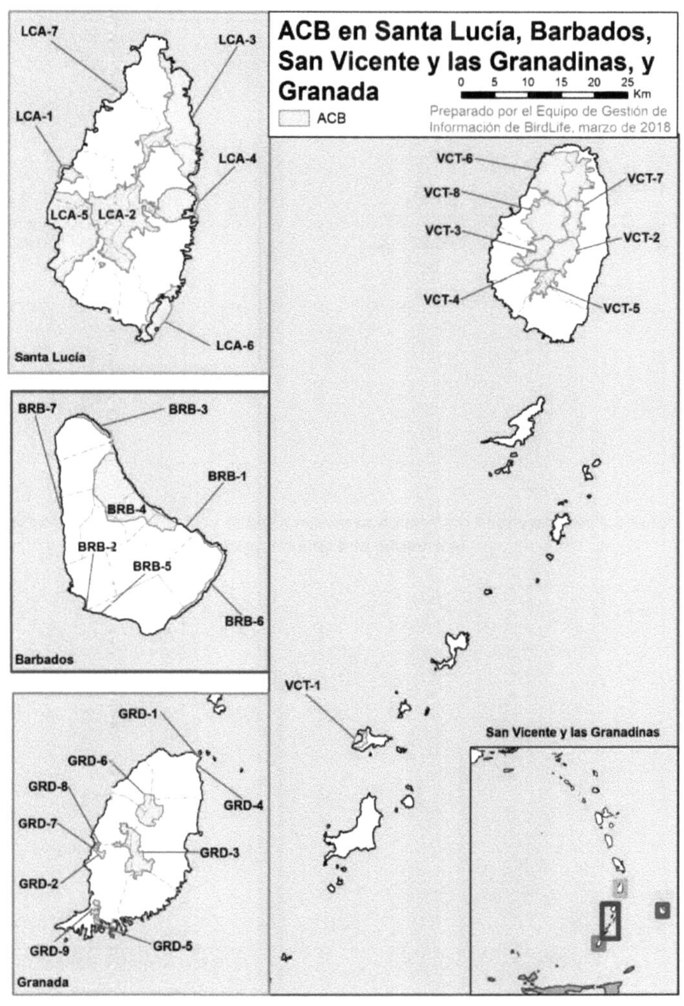

Figure 5.8 Site results in the Dutch Overseas Territories (Sint Maarten, St. Eustatius, Saba, Aruba, Curaçao and Bonaire)

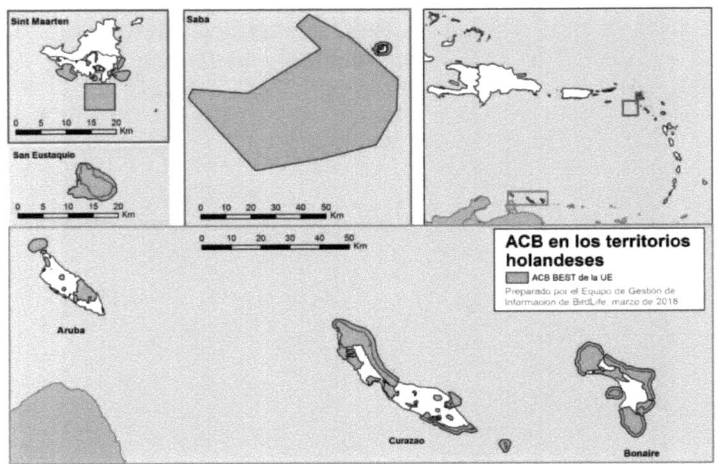

Figure 5.9 Site results in the French overseas regions and territories (St. Martin Martinique and Guadeloupe)

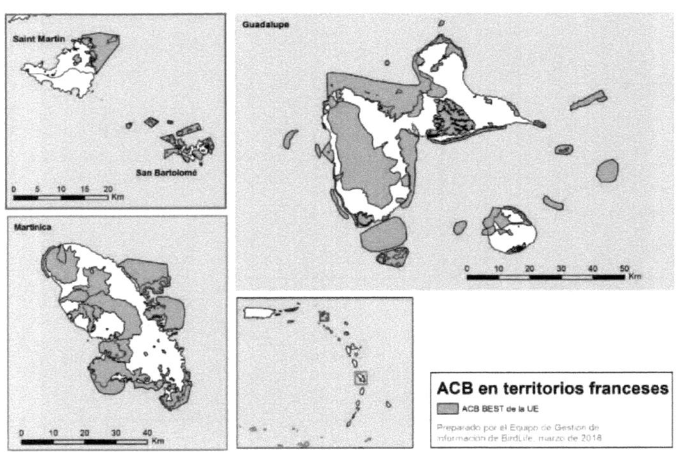

Figure 5.10 Site results in the United Kingdom's overseas territories (Cayman Islands, British Virgin Islands, Anguilla, and Montserrat)[15]

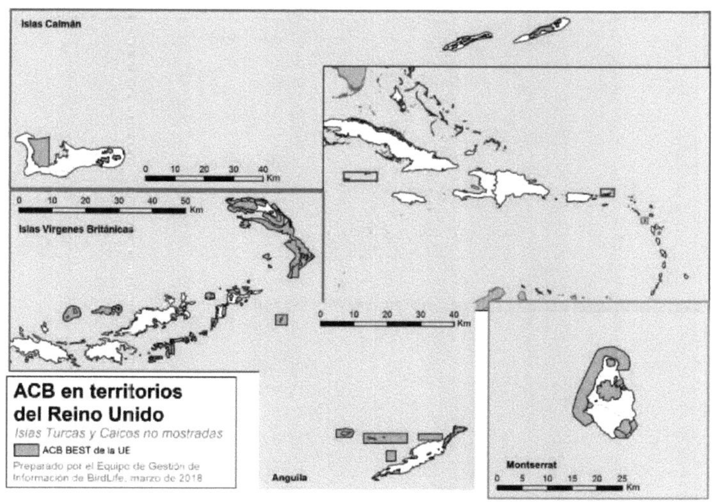

Figure 5.11 Cuba site results

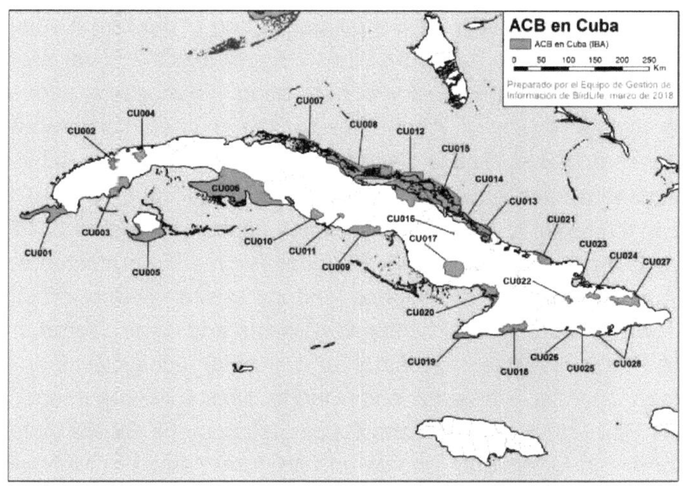

Figure 5.12 Site results in Puerto Rico and the U.S. Virgin Islands

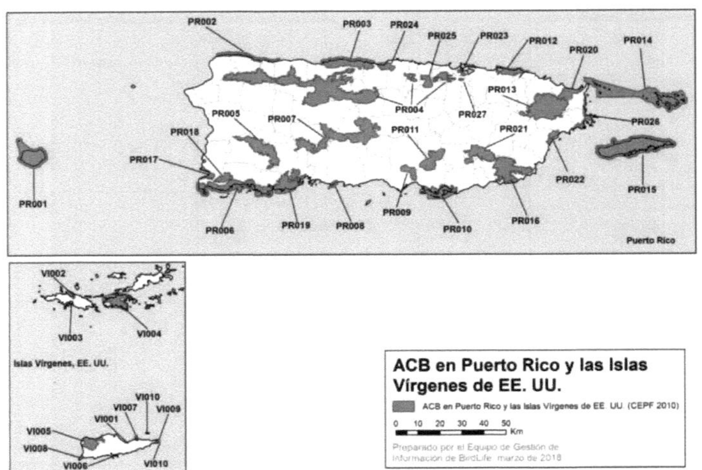

As described in Section 5.2, the description of EU BEST CBAs was based on previous CBA criteria (Langhammer *et al.,* 2007) and, as such, these CBAs follow the actual distribution of the target species, and consider that the habitats of these trigger species cover an area that could be manageable for conservation. Thus, some sites that share the same list of trigger species have been considered as *disconnected* CBAs. These dispersed sites can be considered viable management units on small islands. This is the case for 34 CBAs listed in the regional ecosystem profile (Vaslet and Renoux, 2016).

These CBAs include: a network of lagoons or marine protected areas in Anguilla; mangroves, lagoons and proposed marine protected areas in Aruba; terrestrial habitats in Tortola and Virgin Gorda in the British Virgin Islands; terrestrial and marine ecosystems in the Cayman Islands; terrestrial and marine sites in Guadeloupe, St. Martin, Sint Maarten, Turks and Caicos Islands and St. Barthelemy; dry forests in Martinique; *ghauts and* dry forest areas in Montserrat; and forests in Saba and St. Eustatius. Ecological corridors have been identified between terrestrial, coastal *and* marine CBAs. A total of 43 ecological corridors, covering an area of 2 720 km^2 , are highlighted in the regional ecosystem profile (Vaslet and Renoux, 2016).

Regional ecological corridors were determined according to the migration patterns of bird species, sea turtles, bats and marine mammals and the dispersal of larval marine species (corals, fish, marine invertebrates).

Figures 5.8 to 5.10 show the location of CBAs identified through the EU BEST initiative in the Caribbean region. More information on these sites can be found in Appendix 3.2 and the regional ecosystem profile.

4. THREATS TO WILDLIFE IN THE HOT SPOT

The fauna of the Caribbean hotspot is unique and vulnerable. As in most other island systems, the relative isolation of the islands has resulted in many endemic forms and a biodiversity characterized by small populations, narrow genetic base, reduced competitive ability, limited dispersal opportunities, and restricted distribution. As a result, the native biodiversity of the West Indies hotspot has limited capacity to buffer natural and anthropomorphic threats compared to the mainland biota. However, the region's native biodiversity and ecosystems have evolved in an environment that is affected by regular natural hazards, particularly hurricanes, and therefore have developed some level of natural resilience to disturbances.

Information on threats to biodiversity, their root causes and barriers to effective conservation in the insular Caribbean is scattered and few reviews are available at the regional level. The most recent are those of Brown *et al.* (2007) for the entire region, Vaslet and Renoux (2016) for the EU Overseas Countries and Territories and outermost regions. In many cases, statistics for the Caribbean are grouped with those for Latin or Central America, or are presented for the Caribbean as a whole, which includes the bordering continental countries: Trinidad and Tobago, Guyana and Suriname. National descriptions of threats, root causes and barriers are presented in national biodiversity strategies and action plans (NBSAPs), although, along with other relevant publications, these vary considerably in their presentation of quantitative data and degree of analysis.

Most sources present only qualitative descriptions of hazards and their impacts, causes and measures to address them.

This chapter provides an overview of the main threats to wildlife, biodiversity and ecosystems in the hotspot and their impacts. This is followed by an analysis of the root causes of the threats and the barriers that must be overcome to achieve more effective conservation and sustainable use of the biodiversity and ecosystems of the hotspot, and thus address the threats.

4.1 Threats

Terrestrial biodiversity in the hotspot has been affected by humans since the first arrival of people in the Caribbean some 6,000 to 7,000 years ago. However, negative impacts increased substantially after the arrival of Europeans beginning in the late 15th century, and have increased in the last 50 years due to rapid population growth and island economies in the region (Brooks *et al.*, 2002).

The key threats to fauna and biodiversity presented in this chapter have been prioritized based on quantitative threat data for globally threatened species compiled for the IUCN Red List, updating the subjective prioritization done for the 'Antilles Biodiversity Hotspot Ecosystem Profile 2009'.

The top priority threats to the terrestrial biodiversity of the insular Caribbean, based on a review of threats to the 992 globally threatened species of the hotspot according to the IUCN Red List, are: overexploitation of biological resources; habitat destruction and fragmentation due to agriculture/aquaculture, urban, tourism, and industrial/commercial development; predation and competition from invasive alien species (and other problematic species); and, increasingly, climate change/severe weather events (IUCN 2017b; Table 6.1).

Pollution is a major threat to the marine environment in the hotspot (CEP, 2003). While pollution and sedimentation pose a threat to freshwater ecosystems, they also affect the marine environment.

Given the relatively small size of most of the West Indies, pollution from land-based sources tends to end up in coastal waters. Sedimentation and pollutants flowing downstream affect coastal water quality, smother corals, kill fish, and reduce the tourism and recreational value of beaches in many countries.

Recurring blooms of pelagic algal blooms *Sargassum natans and S. fluitans* have become an increasing threat to coastal and marine ecosystems since 2011.

Sargassum has created immediate problems for tourism and fisheries and has the potential to transport invasive species from one island to another. However, the magnitude of the ecological impact of the blooms is unknown (Franks *et al.*, 2016).

Table 6.1: Threats prioritized for the 992 globally threatened species of the West

Indies hotspot.

Range	Threats	No. of threatened species a world level affected	Percentage of endangered species to world level affected
1.	Overexploitation of biological resources	284,0	29,0
2.	Agriculture/aquaculture	273,0	28,0
3.	Invasive/problematic species	191,0	19,0
4.	Residential/commercial development	168,0	17,0
5.	Climate change/severe weather	88,0	9,0
6.	Human intrusions/disturbances	81,0	8,0
7.	Contamination	49,0	5,0
8.	Energy production/mining	40,0	4,0
9.	Transportation/service corridors	39,0	4,0
10.	Modifications of the natural system	28,0	3,0
11.	Geological events	11,0	1,0
12.	Others	2,0	0,0

4.1.1 Use of biological resources: overexploitation, persecution and control

Unsustained use of the limited and often diminishing biological resources is the main threat to biodiversity (at the species and site level) in the West Indies hotspot. Unsustained use has been identified as a threat to 29% of the globally threatened species in the hotspot (Table 6.1).

IUCN defines threats from the use of biological resources as those related to the "consumptive use of 'wild' biological resources, including the effects of deliberate and unintentional collection; also persecution or control of specific species" (IUCN n.d., p6). The main activities that fall into this threat category in the hotspot include: timber extraction; excessive collection of firewood (especially charcoal); collection of plants for horticulture; unsustainable hunting and egg collection for food or sport; and capture of animals for the pet and aquarium trade. The list of species suffering from unsustainable use is almost certainly conservative, as quantitative data on many of these activities are scarce. This is partly because exploitation is often illegal and therefore hidden, and partly because there is inadequate monitoring due to lack of resources within the relevant environmental agencies.

Timber extraction

The forests of the hot spot have the legacy of past timber harvesting activities. Originally, the hardwood was used to make boats, houses and furniture for early settlers. The rest of the forest was treated as a source of firewood and then logged to make room for plantations.

Today, few islands have significant primary forest cover, and several species that were once common and traded on a large scale are now commercially depleted. These include Caribbean mahogany *(Swietenia mahagoni), which is* non-existent in parts of its former range, and stands of mature forest have been virtually eliminated. It is now listed Endangered and international trade is restricted under CITES. Because of its value, this species was introduced elsewhere and has now naturalized on many islands.

Other economically valuable timber species in the West Indies include West Indian walnut *(Juglans jamaicensis* - VU), West Indian ebony *(Brya ebenus)* and poui oak *(Tabebuia heterophylla).* Illegal logging threatens commercial forest concessions and critical protected areas and buffer zones.

Sixty-six percent of the tropical dry forest in the Caribbean has been transformed into human-dominated landscapes (IPBES 2018). Most of the remaining forest in the Greater Antilles is secondary and only montane forests are relatively intact.

Similarly, in the Lesser Antilles, the best-preserved forest areas are often found on higher ground, for example, in Martinique, Dominica and St. Lucia, or on steep slopes and deep ravines inaccessible for cultivation.

The area covered by forests has increased in some countries in recent decades (Box 6.1 and Section 7.3.3). One such country is Cuba, which was more than 90% forested in 1492 but, by 1900, had a forest cover of only 5%. However, the forest area has increased since then, reaching 13.5 % in 1960 and 30 % in 2015, due to reforestation (FAO

2015).

Box 4.1: Forest cover and deforestation in the Caribbean

Global statistics on forest cover are compiled by FAO every 10 years. The most recent figures (FAO 2015) indicate that most of the remaining hotspot forests are in Cuba (3,200,000 ha), Dominican Republic (1,983,000 ha), Bahamas (515,000 ha), Puerto Rico (496,000 ha), and Jamaica (395,000 ha).

In the Lesser Antilles, regionally significant forest holdings exist in Guadeloupe (71 000 ha), Dominica (43 000 ha) and Martinique (49 000 ha), although, as in the case of the Greater Antilles, the best-preserved areas are in the higher elevations, which tend to be less accessible. The total forest cover of the insular Caribbean amounts to 7 195 000 hectares, or 32 % of the land area (FAO 2015).

FAO figures show that forest cover continues to decline in some of the hotspot countries (particularly Haiti and Jamaica), is stable in others (particularly the Leeward Islands), and is increasing in a few (Cuba, Dominican Republic, Puerto Rico, and St. Vincent and the Grenadines). However, these conclusions should be treated with caution, as there are differences among authorities as to what constitutes a forest and there are no reliable monitoring systems in most Caribbean countries.

The reported figures should therefore be treated as estimates. For example, the Jamaican Department of Forestry has published work that disputes the FAO figure and argues that the rate of forest loss in Jamaica during the 1990s was negligible (Evelyn and Camirand 2003). Similarly, in the Dominican Republic, there is a significant discrepancy between the FAO figures for forest cover (41%) and the government figure (25%).

Most forestry in the region still has a traditional focus on timber production from plantations (often exotic species) and watershed protection, although investment and capacity building by international agencies in recent years is helping to move the sector towards a multiple-use approach, including the protection of natural forests for other ecosystem services such as nature-based tourism and recreation. There is also a trend towards decentralization and, increasingly, stakeholder participation has become an important element of forest management strategies, for example in Jamaica, where local forest management committees have played a role in forest conservation at the local level (Brown and Bennett 2010).

There has been very limited development of forest certification schemes in the insular Caribbean (ITTO, 2008). The only Forest Stewardship Council-certified forest in the West Indies hotspot (365 ha) is in the Dominican Republic (FSC 2017).

Firewood collection and charcoal production

Because energy infrastructure in rural areas of the poorest hotspot countries is still inadequate, communities in these areas rely heavily on fuelwood and charcoal extracted from forested areas, including mangroves. Charcoal and fuelwood provide approximately 70-85% of Haiti's energy consumption (ESMAP 2007 cited in UNEP 2016c) and 80% of the wood extracted in Jamaica is ultimately consumed as fuelwood (FAO 2001).

Addressing the lack of energy sources for poorer rural communities can help reduce demand for fuelwood and reduce pressure on remaining forests and their threatened biodiversity.

In the Dominican Republic, for example, a government policy of subsidizing propane gas and cookstoves was implemented in the mid-1980s, which helped reduce the consumption of wood for charcoal (used for cooking by the majority of the population) from 1 596 000 bags in 1982 to 26 465 bags in 2000 (Gómez and Díaz, 2001). More recently, efforts have been made to promote energy-saving wood stoves (lorena stoves) in the Dominican Republic.

In some countries, such as Haiti, mangrove cutting for charcoal and firewood has become more common as more traditional and accessible wood reserves are depleted. Mangrove loss has implications for coastal resilience and fisheries. Between 1990 and 2000, most hotspot countries showed a decline in mangrove cover (FAO 2007).

Large areas of mangrove have been lost in Martinique, Guadeloupe, the former Netherlands Antilles and the British Virgin Islands. However, since 2000, the mangrove area in the region has remained stable (FAO, 2015). Countries that still have significant mangrove areas are Antigua and Barbuda, Cuba, and the Turks and Caicos Islands.

NTFP Collection
Other non-timber forest products, such as fruits, fibers, resins, tannins, essential oils, tree seeds, honey, fodder, yams and bean poles, ornamental plants, tree fern trunks (for orchid cultivation), bamboo, medicinal plants, spices, edible oils, dyes, gums and mushrooms, which are an important part of the rural economy, especially for the poorest families. However, their value (social and economic) has not been quantified and has only been partially documented in some countries, for example, Cuba and the Windward Islands (John, 2005).

Cuba, for example, listed the production of 1 474 tons of raw material for medicinal and aromatic products, 68 tons of raw material for dyes and colorants, and 18 400 tons of other non-edible animal products harvested from its forests in 2005 (FAO, 2006a). Some NTFPs are known to be harvested at unsustainable levels or using destructive practices.

Hunting
Many animals are hunted for food or sport in the region. Prey hunted for food include many endangered species of amphibians, reptiles, mammals and birds.

Hunted amphibians include the globally threatened mountain chicken, which is found in Dominica and Montserrat. Reptiles hunted include iguanas (Haiti, Dominican Republic, and Lesser Antilles) and sea turtles (especially adult females and eggs), even though international trade in certain turtle species is prohibited by CITES.

The hutias, which were part of the diet of the pre-Columbian inhabitants of the region, are still among the mammals hunted for food.

Several species of birds are also hunted for food, particularly waterfowl and game birds, including endangered species such as the black-billed yaguasa, suirirí yaguaza or West Indian whistling duck *(Dendrocygna arborea* - VU).

Hunting birds for sport, especially pigeons, such as the white-winged dove *(Zenaida asiatica) and* the Caribbean zenaida or Zenaida turtledove (Z. *aurita),* is popular on many islands.

Some target species that can be legally hunted in some countries are becoming increasingly scarce, such as the collared dove *(Patagioenas leucocephala),* which is found in several hotspot countries and is now listed as Near Threatened.

Addressing unsustainable hunting is identified as a conservation objective in the NBSAPs of several countries, either directly (e.g. Jamaica) or through general measures to promote sustainable use of natural resources.

Population censuses of some target species are conducted by environmental agencies on most islands, although most of this information remains in unpublished technical reports.

Hunting seasons and hunting limits are regulated by national legislation, with penalties for infractions. Hunting is generally restricted in formal protected areas, although poaching is still widespread in many of them.

Knowledge and awareness of the threatened status of some hunted species and the importance of associated high biodiversity sites have improved through specific projects, such as the *West Indies Black-billed Whistling Duck and Wetlands Conservation Project* (https://www.birdscaribbean.org/caribbean-birds/wiwd-and-wetlands-conservation-project/#:~:text=To%20reverse%20its%20decline%20and,WIWD%20and%20Wetlan ds%20Co nservation%20program). However, monitoring and enforcement remain key challenges due to lack of capacity and resources among relevant government entities.

There is a general lack of transparency and accurate information on the number and location of animals harvested, as well as on the level of illegal hunting and the impact of hunting on populations. Such information is needed to make informed decisions on species-specific hunting limits, design effective management plans for target species, and protect the most vulnerable species. This represents a major gap in knowledge and research.

Collection of eggs for medicinal use
Seabird colonies in cays throughout the Caribbean have also traditionally been harvested by fishermen during the breeding season for eggs. Although most colonies are now protected by national legislation, illegal egg harvesting still occurs. In Jamaica, human predation has been identified as the most important historical factor contributing to the decline of the country's seabird populations (Haynes Sutton, 2009).

On Hispaniola, the colony of the least tern *(Onychoprion fuscatus)* on Isla Alto Velo was estimated at 175 000 pairs in 1950, but by 1980 it had declined to 40 000 to 50 000 pairs, which is explained by systematic large-scale egg collection by people (Keith 2009).

In St. Vincent and the Grenadines, and Grenada, large-scale poaching of eggs was practiced in the past. Egg collecting was still occurring on the islets of Grenada in the early 1990s and still occurs in the Grenadines (Frost *et al.*, 2009).

Sea turtle egg collection is intensive and widespread throughout the hotspot, although much less so than in Central America. This exploitation and the resulting trade is proving to be a serious management challenge. Some islands report egg poaching levels approaching 100% on some beaches. Exploitation is largely unquantified, and its impact on turtle populations is impossible to judge (Brautigam and Eckert, 2006).

Some endangered or endemic animals are also hunted or collected for medical use. These include the cúa or cuco picogordo *(Coccyzus rufigularis)* in Hispaniola and the boa nebulosa *(Boa constrictor nebulosus)* in Dominica, whose fat is used to make "snake oil," which is believed to help cure joint problems and back pain.

The Cuban crocodile *(Crocodylus rhombifer) and the* American crocodile are captured for their skins. There is some cultivation of the latter species in Cuba, in an effort to address the trade problem with a reintroduction program (Jenkins *et al.*, 2004).

In several West Indies medicinal oil is extracted from leatherback turtles (J. Horrocks *in litteris. ,2009)*.

The economic value of hunting and gathering in the West Indies has not been adequately investigated. Such information would help persuade politicians and other decision makers of the need to increase resources to manage populations of hunted species in a sustainable manner. Some limited data are available for some islands and species, but the picture is very incomplete.

Collecting live animals and plants for trade
Collection to supply local and international pet stores, aquariums and horticultural companies is also believed to pose a direct threat to some species in the hotspot, particularly the more attractive and rare (and therefore more commercially valuable) species, such as parrots and iguanas. Trade statistics for local markets are generally not available and protected species tend to be sold clandestinely. In addition, not all countries in the hotspot submit annual reports on trade in endangered species (UNEP, 2002). As a result, national and international trade statistics for animals and plants are not comprehensive for the Caribbean.

Despite protection under national and international legislation, a small number of threatened species continue to appear in markets outside the region. For example, several specimens of the St. Lucia amazon and Cuban amazon *(Amazona leucocephala)* have been reported in EU countries in recent years, despite both species being listed in EU Annex A and CITES Appendix I (Anon., 2002 cited in Theile *et al.*, 2004).

In 2011, 74 eggs of the black-billed amazon (*A. agilis* - VU) and yellow-billed amazon (*A. collaria* - VU) were seized at Vienna International Airport, Austria. Forty-five chicks hatched from these eggs, after successful incubation at a zoo in Vienna (Ferguson, 2011).

Illegal trade in the overseas territories of the United States, United Kingdom, France and the Netherlands in the region is also of concern, although the extent of illegal

In general, more comprehensive surveys are needed to quantify current levels of exploitation of animals and plants in the hotspot and regional and national reviews of their use. There is also a need to establish scientific limits for exploitation of target species, as well as better monitoring and awareness programs for target species, more clearly defined laws and regulations for the use of wildlife species, and improved law enforcement. Finally, national and regional reporting, including CITES reporting, needs to be improved.

Fisheries
Caribbean fishery resources are among the most overexploited in the world, and regional production has declined by more than 40 % in the last two decades (FAO 2014).
Although small in global terms, Caribbean fisheries are important to the livelihoods of coastal communities and food security. Most Caribbean fisheries are artisanal and although fishing practices have changed little, apart from the introduction of motorized boats and modern gear, what has changed is the level of intensity and the number of people making a living from fishing (Johannes, 1977 in Hawkins and Roberts, 2004; Polunin *et al.*, 1996 in Hawkins and Roberts, 2004).
Fifty-four percent of species or species groups in the Caribbean are considered overfished or fully depleted (Western Central Atlantic Fishery Commission, 2017). Overexploitation is the main threat to bony fishes in the Caribbean; it directly affects half of the species in the Wider Caribbean listed by the IUCN as globally threatened or near threatened (Linardich *et al.*, 2017).
Artisanal fisheries have transformed coral reefs in ways that seriously compromise their ecological and economic value (Hawkins and Roberts, 2004). Fishing pressure has been shown to have a cascading effect on coral reefs, with overfishing contributing to changes in the structure and composition of Caribbean coral reefs and increasing the abundance of algae and sponges due to declining populations of herbivores, such as parrotfish (Hawkins and Roberts, 2004, Loh *et al.*, 2015). More than half of the overfished species in the Wider Caribbean are found on reefs (Linardich *et al.*, 2017).

4.1.2 Agricultural and aquaculture expansion and intensification

Expansion and intensification of agriculture and aquaculture is an identified threat to 28 % of all globally threatened species in the West Indies hotspot (IUCN, 2017b).

Large-scale land clearing for agriculture, mainly sugarcane plantations at lower elevations, began in the 16th century, shortly after European colonization began, and increased during the 18th and 19th centuries, leading to widespread deforestation throughout the region (timber is used for construction and fuel in the sugar mills). This led to destabilizing erosion, loss of some permanent streams and decreased soil fertility (McElroy *et al.,* 1990).

Soil is crucial for maintaining both biodiversity and the ecological services it provides, but this important resource continues to degrade globally (United Nations Convention to Combat Desertification, 2017). Some of the smaller islands, such as Antigua, Barbados, Bahamas, Bonaire, St. Kitts and Nevis, and the U.S. Virgin Islands, lost virtually all of their native forests at that time or were completely altered by agricultural development.

The subsequent increase in new agricultural export markets led to new periods of intense deforestation, such as during and after the banana boom of the 1970s and 1980s in the Windward Islands. Recent agricultural threats to mountain forests come from the spread of cocoa, coffee and tobacco plantations. The abandonment of sugar (and other important crops, such as cotton, on some islands) due to changing economic conditions or reduced soil fertility often resulted in conversion to pasture and a large increase in livestock production.

Overgrazing has significantly altered the vegetation of many forested areas, resulting in degraded scrub vegetation, and continues to be a threat to native vegetation, especially on those islands with significant numbers of sheep and goats, e.g., Bonaire, Grenada's Carriacou, and Petit Martinique, St. Barthelemy and many offshore cays that have traditionally been used as natural goat corrals.

Agricultural expansion has resulted in unsustainable levels of cultivation and grazing on unsuitable land (Rojas *et al.,* 1988), leading to soil erosion, further land degradation and landslides that cause significant economic losses each year and are especially damaging in rugged islands with coastal plains, such as Jamaica and Hispaniola.

Most of the Antillean forests have been lost due to agricultural development. Today, no more than about 21 600 km^2 or less than 10 % of the original vegetation remains in a pristine state in the West Indies hotspot (FAO, 2015).

Cuba has the largest remaining tracts of forest in the Caribbean, but these still represent only 30% of the original area and a significant portion is reforested land (FAO, 2015).

While clearing for agriculture has been one of the greatest threats to native forests in the insular Caribbean, the decline in some agricultural markets has led to the abandonment of degraded areas and the subsequent expansion of secondary forests, which often still have high biodiversity and ecosystem service values.

Natural succession has been slowly reforesting Puerto Rico since the 1950s, when government policies promoted industry over agriculture. Natural forest cover has expanded from a minimum of 6 % of the island in the 1940s to 58 % by 2015, due to reforestation and the abandonment of shade coffee plantations (Helmer *et al.* 2002, FAO, 2015).

The cessation of traditional agricultural activities on Curaçao, due to economic factors, the theft of livestock and agricultural products and the increase in the speculative value of private property, has led to a new growth of dense secondary forests, especially in the western half of the island.

In the U.S. Virgin Islands, secondary forest has regenerated after the end of large agricultural plantations.

Secondary forests provide important ecosystem services. Watershed protection, water supply and fuelwood provision are especially important in the Antilles. These forests could provide important opportunities for carbon sequestration as part of climate change adaptation and mitigation strategies. However, to date, forest conservation efforts in the hotspot have largely focused on the remaining areas of primary forest, while secondary forests and abandoned agricultural areas often appeal to development planners as being of lesser importance (Massol Gonzalez *et al.,* 2006).

Agricultural practices that degrade the environment or increase pressure on already scarce resources are common in the region and are the root cause of much of the erosion, pollution and sedimentation that threaten both marine and terrestrial environments and can also increase the likelihood of fires (Burke and Maidens, 2004).

Extensive areas of freshwater wetland habitat in the Caribbean, such as marshes and lagoons, have also been drained and converted to agricultural schemes, or degraded due to overgrazing by livestock.

There are no recent and accurate figures for the area of wetlands lost. A 1990 survey of coastal wetlands (predominantly mangroves) on 16 islands in eastern ntilles revealed evidence of damage at almost all sites visited, with over 50% reporting severe damage (Bacon, 1993). Uncontrolled aquaculture development has also led to the loss and degradation of wetland habitats in some coastal areas, including coastal lagoons and mangroves in certain countries.

Poor siting, construction and operation of ponds in areas exposed to storms and flooding can introduce exotic species and diseases into ponds and other habitats in coastal ecosystems. Exotic fish species, such as claria, Nile tilapia *(Oreochromis niloticus)* and trout *(Salmo* spp.), for example, have been accidentally or intentionally introduced into streams, ponds and wetland areas through aquaculture projects, where they compete with native fish populations (CEC, 2001; Kairo *et al.,* 2003; FAO, 2006c).

4.1.3 Invasive and other problematic species, genes and infectious diseases

The establishment and spread of IAS (and other problematic species, genes and d seases) have been implicated in extinction risk in the Caribbean (IPBES, 2018).

IAS pose a threat to 19% of the globally threatened species of the hotspot, especially endemic species (IUCN, 2017b). The most damaging on islands are typically terrestrial vertebrates, such as goats, feral cats, pigs and rats. These species are responsible for more than half of all animal extinctions on islands globally.

Invasive alien species contribute to extinction risks to the greatest extent in North America, followed by the West Indies.

As in other islands, Caribbean habitats are vulnerable to the impacts of invasive species due to the generally reduced populations of native species, the evolutionary effects of isolation, and the release of introduced species without natural enemies (Kairo *et al.,* 2003).

The spread of IAS is facilitated in the Antilles by the region's dependence on imports, the high degree of exposure to extreme weather events and the multiplicity of pathways that exotic species can use to reach the islands.

Historically, many species have been introduced deliberately or accidentally, and this process has continued to the present. In many places, these populations have persisted, causing constant devastation on the islands.

The potential for deliberate or accidental introduction of other IAS, such as marine species and agricultural pests, has increased in recent years through globalization and the associated increase in international trade, tourism and transport links. In addition, changes and development in some sectors, especially agriculture and aquaculture, have provided opportunities for the introduction and spread of IAS.

The risk of accidental introduction of diseases and pathogens and the potential escape of traded species into the wild constitute a "hidden cost" of international trade, posing a serious risk to the economic and ecological health of all hotspot nations.

According to the Federal Maritime Commission, trade and transport between the southeastern United States, the Caribbean and Latin America could triple from 2005 levels by 2020 as a direct result of the Dominican Republic-Central America Free Trade Agreement.

While the catastrophic impact of invasive terrestrial vertebrates on islands is well known, no experimental studies have been conducted to verify the exact impacts for most other IAS.

The introduction of IAS and their effects are phenomena that are currently occurring in the hotspot and countries face new and emerging threats. For example, the invasive seagrass species *Halophila Stipulacea* has spread rapidly throughout the eastern Caribbean since 2002 and is a potential threat to the functioning of local seagrass ecosystems (Smulders *et al.,* 2017).

Now that they are as far north as the U.S. Virgin Islands and Puerto Rico, the marine conservation implications of this IAS appear serious and are currently under study.

A review of invasive species threats in the West Indies region identified 522 alien species, including 449 terrestrial (390 naturalized/invasive), 55 freshwater (10 naturalized/invasive) and 18 marine (16 naturalized/invasive) (Kairo *et al.,* 2003).

Introduced terrestrial species vastly outnumber introduced marine and freshwater species, although this is probably a reflection of insufficient sampling of the marine environment (Kairo *et al.*, 2003). Numbers on individual islands can be very high. For example, 138 species have been reported as invasive in the Dominican Republic (Ministry of Environment and Natural Resources, 2011b), including 17 of the world's 100 worst invasive species (Lowe *et al.*, 2001), and in the Cayman Islands over 100 alien species of flora and fauna have been reported (JNCC, 2007).

Current information on species known to be naturalized or invasive in the insular Caribbean can be found in the IUCN Invasive Species Specialist Group's 'Global Invasive Species Database'. Available at www.issg.org

The main pathways for the introduction of IAS into the West Indies include trade in agricultural products, the pet and aquarium trade, ill-conceived biological control schemes, agricultural, forestry and aquaculture development projects, and horticulture. The latter is of particular concern and has a strong connection to the tourism sector through horticultural imports.

In most of the West Indies, live plants comprise the majority of nursery imports, which largely come from the United States. In Anguilla, for example, the Cuban tree frog *(Osteopilus Septentriionalis) and the* giant African snail *(Achatina fulica)* were probably introduced through the importation of containers of exotic plants or other construction materials to supply the developing tourist industry. The snail has also been reported in the Dominican Republic.

Phytosanitary standards are governed by the 1952 International Plant Protection Convention, and all Caribbean island countries are signatories or adherents to the convention. However, border biosecurity is relatively lax, and the sheer volume would overwhelm existing quarantine measures. Little documentation is available on invasive species interceptions at ports.

Given that the tourism sector is the largest consumer of imported horticultural products in the entire region, there is a need for proactive engagement of the tourism sector (architects, landscape architects, contractors, gardeners and nursery owners, as well as national inspection and quarantine services) to agree on measures to reduce the risk of IAS introduction. Locally, areas abandoned after slash-and-burn agriculture do not regenerate quickly with native vegetation, but are colonized by invasive plant species. Fortunately, for the most damaging IAS (invasive terrestrial vertebrates), well-developed biosecurity systems exist and, when implemented, reintroduction, accidental or intentional, to islands where these species have been eradicated is rare.

Regional initiatives and groups seeking to address IAS in the West Indies include the Caribbean Regional Invasive Species Intervention Strategy (CRISIS) and the Caribbean Invasive Species Working Group (but dealing primarily with agricultural pests).

In 2013, the International Coral Reef Initiative spearheaded the development of a regional strategy to guide action to control invasive lionfish *(Pterois volitan)* in the Wider Caribbean.

A major GEF-funded regional project, *Mitigating the Threats of Invasive Alien Species in the Insular Caribbean,* implemented by UNEP and executed by CABI with a variety of national, regional and international partners, ended in 2014 after focusing on: developing national strategies; establishing Caribbean-wide cooperation and strategies; improving information management; preventing introductions; and early detection. However, there has been little follow-up at the regional level and a major programmatic gap in the Caribbean remains the practical eradication of IAS on a large scale, based on the experience of the few eradication projects that have been implemented.

At the national level, most countries in the region have identified IAS as a major threat to their biodiversity and highlight the need for control activities as a primary target in their NBSAPs or in their National Reports to the CBD. The Bahamas, for example, has a National Invasive Species Strategy, which was updated through the GEF project mentioned above; Jamaica's NBSAP describes 45 specific objectives related to IAS, where the preparation of an IAS management strategy is indicated as a key priority.

Some countries, such as the Dominican Republic, have developed national strategies for the control and eradication of invasive species (Ministerio de Medio Ambiente y Recursos Naturales, 2011b).

Quantitative data on Caribbean invasive species (numbers, distribution and impacts) is considered inadequate and limits the ability to design effective responses.

Lack of information remains a weakness that needs to be addressed (Kairo *et al.,* 2003). There is also a low level of awareness, from the general public to policy makers, of the threats posed by IAS and their environmental and economic impacts.

A particular challenge in addressing IAS arises from the fact that many of the main pathways for species introductions (related to trade and tourism, for example) are critical to national economies. Emerging infectious diseases are a recently recognized threat to biodiversity globally and in the Caribbean. Chytridiomycosis in amphibians is a notable example of this threat (Daszak *et al.,* 2000). Caused by the recently described chytrid fungus *Batrachochytrium dendrobatidis*, chytridiomycosis is a disease that can lead to the extinction of amphibian populations and species (Skerratt *et al.,* 2007; Chenga *et al.,* 2011).

More than 200 species of frogs and salamanders are known to be susceptible to infection. Population declines due to the disease have occurred in Australia, America and Europe (Berger *et al.,* 1998; Lips *et al.,* 2006; Bosch and Rincon, 2008).

In many of the 122 amphibian species extinctions that have occurred since 1980, particularly those where species have disappeared from pristine areas, chytridiomycosis is suspected to be the primary cause (Skerratt *et al.,* 2007 and IUCN, 2008).

In the Caribbean, amphibian chytrid fungus is known to occur on the islands of Puerto Rico, Hispaniola, Dominica, Cuba and Montserrat. The disease has been implicated in the decline of mountain chickens in Dominica and Montserrat, and the possible extinction of three species in Puerto Rico is suspected (Burrowes *et al.,* 2004 and Díaz *et al.,* 2007).

Chytridiomycosis presents a particular challenge for wildlife conservation because the routes of transmission and how it kills amphibians are not well understood.

It is believed to be transmitted by the introduction of infected animals, water, vegetation or soil to a new region. In addition, species are affected differently by the disease: it is very lethal to some species, such as the mountain chicken, while others may harbor lethal infections and spread the fungus to sensitive or highly susceptible species. Therefore, habitat protection is necessary but not sufficient to protect many species from chytridiomycosis.

Species-specific conservation actions, in addition to site protection, will be required to protect the amphibian populations most vulnerable to this threat.

4.1.4 Residential, commercial, industrial and tourism development

Habitat loss to residential and commercial development has been identified as a threat to 17% of all globally threatened species in the hotspot (IUCN, 2017b).

The considerable growth of population and economies in most Caribbean countries over the past 50 years has been accompanied by extensive urban industrial and commercial developments and associated infrastructure, which has occurred without adequate planning.

This has led to the destruction and degradation of huge areas of natural habitat, transforming the landscape and character of many of the Antilles. Impacts have included: untreated sewage pollution from residential and tourist developments; industrial pollution; removal of natural coastal vegetation for the construction of housing, hotels, resorts, commercial complexes and roads; removal, dredging, channeling or filling of coastal wetlands (lagoons, estuaries, coastal marshes) and mangroves for ports and harbors; sand mining and erosion of beaches and dunes.

As residential and tourist populations have grown, there has been an increase in water consumption from surface and groundwater sources, contributing to salinization, changes in ecosystem function, and reduced water supply availability. Housing as well as commercial/industrial initiatives have been located on agricultural lands, displacing farmers to more marginal lands.

While these developments have been a major force for economic growth, modernization and improvements in human welfare in the region, they have had negative impacts on the environment.

The major concern has been the huge and uncontrolled growth of tourism in the Caribbean region, with widespread construction of hotels, ports and associated developments, especially along coasts with white sand beaches and offshore coral reefs, often resulting in beach erosion and other profound impacts (UN EP-RCU, 2001 and UNEP, 2004b).

These tend to be leeward beaches with gentle waves: the preferred nesting sites of the remaining populations of Critically Endangered hawksbill turtles.

Development has often meant the complete removal of natural shoreline vegetation, the planting of ornamental trees, shrubs and lawns for golf courses and gardens, the filling in of mangroves for harbor developments and mosquito control, and the construction of new roads to provide access to coastal areas that previously could only be reached on foot or by sea.

Overall figures for the area of natural habitats lost to tourism development in the hotspot are not available, but the total area is considered huge, with very few coastal areas remaining unaffected.

Infrastructure projects, such as road construction, are often inextricably linked to major tourism developments and can have profound effects on biodiversity.

Many tourism sites operate beyond their carrying capacity, both from a biophysical and management standpoint. High tourist flows during the high season, for example, often overload public services, reduce local food and water supplies, and generate large amounts of solid and liquid waste, considering that local municipalities have very limited waste management facilities.

However, some tour operators are taking a more environmentally responsible approach. For example, the Puerto Rico-based Caribbean Hotel and Tourism Association (CHTA) has been a strong supporter of sustainable tourism, particularly through the Caribbean Alliance for Sustainable Tourism (CAST) and through the promotion of the Green Globe and Blue Flag certification programs.

In collaboration with the Barbados-based Caribbean Tourism Organization (CTO), the region's leading tourism trade organization, CHTA has developed a position paper on Caribbean tourism and climate change and supports an initiative to make Caribbean tourism carbon neutral.

4.1.5 Climate change and extreme weather events

While it is accepted that climate change has adversely affected biodiversity at the genetic, species and ecosystem levels, and will continue to do so, there is not a full understanding of the extent of climate changes that are already affecting species and ecosystems in the West Indies hotspot (Section 10.3).

Although climate change has been identified as a threat to only 9 % of threatened species in the IUCN Red List analysis of documented threats (IUCN, 2017b), it is expected that, over time, it will be recognized as a major threat to biodiversity in the hotspot.

Climate change interacts with other threats to increase the vulnerability of species and ecosystems. Mangrove ecosystems, for example, already weakened by conversion to other land uses, are susceptible to impacts such as sea level rise, changing ocean currents and rising temperatures.

4.1.6 Alterations related to human use

The region's increasing human population, expanding agriculture, and urban and tourism developments mean that there are now few relatively pristine natural areas outside of protected areas and inaccessible mountainous regions that are not subject to some form of human disturbance.

Even within protected areas, increased visitor numbers in recent years have led to vegetation degradation and disturbance of wildlife as carrying capacity has been exceeded, such as along the Blue Mountain Peak trail in the Blue and John Crow Mountains National Heritage CBA in Jamaica.

Fire is a major cause of human-induced disturbance in the Caribbean and is commonly used to clear land for agriculture and to make way for human settlements, prepare sugarcane fields for cutting, clear undergrowth in forests, and encourage new growth in grasslands and sparsely wooded areas for grazing (FAO, 2006b).

Forest fires in the insular Caribbean mainly affect dry forests (500 to

1 000 mm mean annual precipitation), but even montane forests with higher precipitation (1 000 mm or more per year) suffer fires in exceptionally dry years (Robbins *et al.*, 2008). Much of the vegetation of the West Indies hotspot (e.g., Jamaica, Puerto Rico and the Lesser Antilles) is not adapted to fire and is adversely affected by fire.

In fact, conservation efforts to protect forests are often thwarted by deliberate forest fires, even within protected areas and forest reserves, to convert them to pasture or agricultural land.

However, the forests of the Bahamas (including the Turks and Caicos Islands), the pine forests on Hispaniola and Cuba, several species of palms that form extensive savannas in Cuba, and some types of herbaceous wetlands and localities on these and other islands, such as the Zapata swamp in Cuba, have evolved to fire and are dependent on fire for existence in their present forms.

Other species depend indirectly on fire. For example, the primary nesting tree of the Cuban parrot is the fire-adapted savanna palm *Colpothrinax wrightii. Consequently,* fire is not only a threat in the region, but a critically important natural process in some systems and an important soil management tool, which has the potential to be managed to minimize its negative aspects or maximize its positive aspects (Myers *et al.*, 2004a, b).

Fire information in the region is limited and, in many cases, non-existent. This represents a major gap in research, given the potential loss of habitat due to fire and the resulting vulnerability to invasive species.

4.1.7 Contamination

The main sources of pollution in the insular Caribbean are sewage and urban effluents (often untreated or undertreated); agricultural pesticides and fertilizers (mainly nitrates, phosphates, pesticides, fungicides, and herbicides from nonpoint sources); spills and accidents involving heavy metals and oils at industrial facilities (also oil from marine sources reaching the beaches); toxic chemicals from mining operations; and solid wastes from various sources.

Eutrophication is also caused by the disposal of large quantities of effluent from sugarcane juice extraction (called *"dunder"* in some islands), which is discharged into drains and rivers. Waste management and disposal capacity (both solid and liquid) is very limited in the insular Caribbean and, as a result, pollution of coastal areas, especially from land-based sources, is a major threat to coastal biodiversity (including mangroves, beaches, and coastal lagoons).

Waste management is a major environmental problem in the Caribbean, where urban population growth, industrial activity and tourism continue to outstrip infrastructural capacity to handle waste, often resulting in disposal in inappropriate landfills.

Riparian and coastal ecosystems are negatively affected by untreated or partially treated wastewater. The use of disposable, non-biodegradable packaging material, such as food containers, aluminum beverage cans, and plastic bags, adds to the solid waste problem.

The expansion of the tourism industry and the increase in the number of cruise ship tourist arrivals have also contributed to the increase in the total amount of solid waste generated in the hotspot.

While most Caribbean countries no longer use some of the most dangerous pesticides, such as DDT, dieldrin and toxaphene, other pesticides can be long-lasting and still pose a threat. For example, in Guadeloupe, chlordecone, an organochlorine-based insecticide, used intensively in the past against weevils in banana plantations and banned since 1993, has permanently poisoned some of Guadeloupe's soils and waters (Belpomme 2007).

Efforts have been made to ensure access to potable water, but the soils of some parts of this and other islands in the Antilles are hopelessly contaminated.

Heavy metal pollution is also a problem in some places, for example, the Salines Lagoon in Martinique. Most of the hotspot countries are signatories to the Stockholm Convention on Persistent Organic Pollutants (POPs), which seeks to protect from chemicals that remain intact in the environment for long periods.See,http://chm.pops.int/Countries/StatusofRatifications/Parti esandSi gnatoires/tabid/4500/Default.aspx

Unfortunately, figures for overall soil and river pollution in the region are not available, due to inadequate monitoring (a reflection of lack of resources) on most islands, and its impact on terrestrial ecosystems and biodiversity is poorly known. It is therefore difficult to assess how serious the pollution is in relation to other threats.

Much more research has focused on the impact of pollution in the marine environment, where municipal, industrial and agricultural waste and runoff account for up to 90% of all marine pollution in the region (CEP 2003, Heileman and Corbin 2006).

A major program to address pollution in the insular Caribbean is UNEP's Caribbean Environment Programme (CEP), which seeks to reduce pollution of the marine environment by improving coastal management and environmental monitoring, promoting sustainable agriculture, improving wastewater treatment and restoring polluted bays. It also aims to develop best management practices for erosion and sediment control, water and land use management, and pesticide and fertilizer control.22 It is also recognized that pollution has significant socioeconomic impacts in the region, including on human health (UNEP 2004a, b).See http://www.cep.unep.org.

4.1.8 Energy production and mining

Mining

There has also been a major loss of natural habitats due to mining activities in some countries. This is most notable in Jamaica, where significant areas have been lost, particularly of native forests in the center of the country, due to bauxite and limestone extraction; also threatening some fairly pristine rainforests in limestone areas.

Bauxite mining has also occurred in Cuba and Hispaniola, although nickel, cobalt, iron and copper are Cuba's main mining products.

Mining industries in the region have a spotty record of complying with requirements to "restore" lands devastated by mining, and governments have an equally poor record of enforcing penalties for non-compliance.

EIAs are often little more than paper exercises in many countries. In addition, restoration attempts have not been very successful in revegetating areas with native species.

Instead, common and widespread weedy species tend to dominate. In addition, given the long history of mining in the region and the continued importance of the mining sector in the national economies of some of the high biodiversity countries, ecological restoration of mining areas remains a priority area of research.

In addition to direct damage, mining activities in the Caribbean have also opened up access to previously remote areas, facilitating the movement of people to these areas (for jobs related to mining activities or support services). In turn, this has led to an increase in small-scale agriculture, especially slash-and-burn agriculture, illegal hunting, firewood collection and charcoal production. Even subway mining operations cause damage, through the clearing of vegetation for surface-level facilities and the dumping of tailings, with the risk of contamination due to poorly constructed or poorly managed effluents and tailings ponds.

There has also been an increase in the illegal extraction of gravel from riverbeds and sand from beaches for the construction of hotels, resorts and residential houses: practices that are common and widespread in the Antilles.

In addition to destroying turtle and seabird nesting habitats and coastal and floral communities, beach sand mining causes sedimentation and disrupts hydrology, which has negative impacts on neighboring coral reefs and other marine ecosystems.

Although their cumulative impact is considered significant, these activities tend to be localized and small-scale, making it difficult to enforce compliance.

The insular Caribbean is heavily dependent on imported oil for energy (90% of all energy used) and there are no significant coal deposits on the islands.

Wind (Barbados), hydroelectric power (Dominica, Dominican Republic and St. Vincent) and solar energy are considered potential alternative energy sources.

The installation of generation systems for these forms of "clean" energy involves a certain amount of habitat loss. In addition, wind farms can pose a threat to bats and birds, as they can be injured and killed by turbine blades. Consequently, the siting of future wind farms is critical and comprehensive EIAs are needed in all cases.

4.1.9 Geological events

There are about 30 active or potentially active volcanoes in the Lesser Antilles. Volcanic activity no longer occurs in the northern part of the region. Major events in the last 100 years have only occurred on the main peaks of Guadeloupe, Martinique, St. Vincent and, more recently, Montserrat.

The 1902 eruption in Martinique was responsible for the extinction of an endemic rodent, the giant Antillean rice rat *(Megalomys demarestii).* After a major eruption, it takes several decades for the vegetation to return to its normal appearance.

Vegetation near permanent active fumaroles and sulfur springs, such as in Montserrat, Dominica and St. Lucia, is adapted and limited to a few sulfur-tolerant genera, such as *Clusia* and *Pitcairnia.*

4.1.10 Threats due to development programs and initiatives in the region

There is a clear need for more integrated and cross-sectoral land-use planning, including tourism, agriculture, forestry, industry, transport, mining, energy and environment, that takes into account all the economic, social and environmental costs and benefits provided by ecosystem services (Section 4.6). Stricter implementation of EIAs, as well as valuation of ecosystem services, is also needed to increase awareness of the value of natural ecosystems among politicians, other decision-makers and planners.

An increasing number of major banks and international financing institutions are adopting the Equator Principles (www.equator-principles.com) which require lending sources to ensure that the projects they finance are developed in a socially responsible manner and reflect sound environmental management practices (comply with international best practices and standards), including rigorous and comprehensive EIAs.

Consequently, more attention should be paid to alerting international financial institutions to the potential threats of the developments they are financing in CBAs.

4.2 Root causes and barriers

4.2.1 Medullar Causes

There is a complex mix of interacting socioeconomic, political, cultural and environmental factors driving environmental change and threatening biodiversity in the insular Caribbean.

The main ones are increasing population and material consumption, poverty and unequal access to resources, the inherent economic and environmental vulnerability of islands to external forces such as changes in global trade regimes and climate change.

Some of these, such as poverty, are local or national problems, while others, such as climate change, require global attention to resolve. All of these drivers can be exacerbated or mitigated by public policies and institutional arrangements at national, regional and international levels.

Population growth and movements
At a fundamental level, many trends affecting biodiversity and ecosystems in the insular Caribbean reflect the limited land available to an increasing number of users.

The West Indies has some of the highest population densities in the world. The regional population in 2016 was approximately 38 million, and is projected to increase to around 46 million by 2030 (Population Reference Bureau, 2017). Some countries, such as Haiti, are expected to face substantial population increases (Section 7.1.2).

All countries are experiencing rapid rates of urbanization and migration from rural to urban areas, resulting in increased demand for natural resources, particularly water, energy and land for construction, and increased problems associated with waste management and sanitation.

These demographic changes have increased the concentration of people in ecologically sensitive areas, especially in coastal zones and mountain slopes, leading to severe environmental degradation in some countries.

The relatively high population densities of the islands also mean that there is the potential for conflict over scarce resources and land, particularly in coastal areas. On the drier islands there are conflicts over water.

Rapid economic growth and increased consumption

Along with population growth, many countries in the region have seen an increase in GDP and average incomes in recent decades, with the rise of a middle class that has generated demand for the goods and lifestyles of the developed world. Along with increased trade, which has increased the incidence and risk of IAS introduction, changing consumption patterns have led to increased pressure on land for housing and urban development, as well as for environmental services, particularly energy and drinking water.

In the case of water, especially the reliable provision of clean water, demand is exceeding the capacity of the natural supply. This is due in part to the enormous demands of the agricultural and tourism sectors and the reduction in supply, quality and reliability as a result of forest conversion and pollution and soil erosion in watersheds.

Agriculture is the largest consumer of water in the Caribbean. For example, it accounts for more than 90% of the total water used in Haiti. The tourism sector also consumes enormous amounts of water, and the low-lying limestone islands of the eastern Caribbean that suffer the highest rates of water scarcity are also among the most attractive for mass tourism.

By international standards, Barbados, Antigua and Barbuda, and St. Kitts and Nevis are already considered water-scarce countries, meaning that they have a water supply below 1,000 m3 per capita per year (UNEP, 2008). Changes in rainfall patterns and pronounced periods of localized drought associated with climate change are expected to increase water stress.

Poverty and inequality

The Antilles are, apart from Haiti, middle- or high-income countries. However, there are high levels of economic inequality in some countries (Section 7.2.3).

Poor people in the Caribbean often depend directly on natural resources, but are often forced to use them unsustainably due to immediate survival needs. Consequently, poverty is considered a root cause of biodiversity and ecosystem loss and degradation in many of the islands. While marginalized groups in Haiti are responsible for some of the country's environmental degradation, Haitian refugees who risk their lives on sea crossings to neighboring countries may be environmental as well as economic or political refugees (Brown *et al.* 2007).

Lack of legal land ownership and access to land and resources are two of the key determinants of poverty in the insular Caribbean. In addition, poor groups and individuals have little voice in decision-making and have fewer rights, and are often displaced or dispossessed by existing power structures and vested interests. Control over natural resources and their use has been, and continues to be, in the hands of the rich and powerful, including governments.

As a result, poor farmers and rural communities have few alternatives other than clearing remaining forests for subsistence crops on marginal lands prone to erosion or overexploitation of natural resources to obtain food and cash essential for their short-term survival.

Lacking technical support, hillside farming practices tend to be poor, resulting in low yields, increased soil erosion and disruption of hydrological systems (demonstrated dramatically in Haiti, although the problem exists throughout the region), which after a short period of time leads to increased demand for land and further clearing of forests and other natural habitats. In addition, the lack of property rights acts as a disincentive to invest in sustainable land management practices.

Because of their dependence on biodiversity and ecosystem services, those most affected by environmental degradation tend to be the rural poor.

Policies and incentives that harm the environment
With the exception of Cuba, Caribbean governments have followed the dominant global economic models, through policies based on export-oriented development, especially for agriculture, and, in recent years, the provision of services, especially in the tourism and financial

sectors. These development policies have generally failed to integrate conservation and resource management considerations in a systematic and participatory manner.

Associated with these policies have been economic incentives/subsidies, subsidies and financial arrangements for favored sectors, such as reduced rates for water and electricity, tax exemptions for investments and exports, subsidized prices for imported fertilizers and pesticides, and construction of transportation and communications infrastructure to facilitate development, all of which have encouraged unsustainable extraction of natural resources and environmental degradation. For example, government policy in many Caribbean countries has been to expand tourism as a means of generating jobs and foreign exchange, actively soliciting outside investment and often granting favorable terms to developers (Section 8.4.2).

In addition to national policies, the policies of some major donors have been criticized for encouraging the multiplication of development projects without taking into account their impact on biodiversity, including the allocation of funds and subsidies to overseas countries and territories and outermost regions (Palasi *et al.*, 2006).

Dependence, isolation and inherent vulnerability

The Antilles, like other SIDS worldwide, share a number of anthropogenic and natural characteristics that make them particularly vulnerable to the impacts of a wide range of internal and external forces that can threaten biodiversity and natural environments and constrain sustainable development (Griffith and Ashe, 1993; Kaly *et al.*, 2002).

Due to their small size, insularity and the characteristics of their natural resource base, most Caribbean countries are dependent on trade and external sources of energy and, consequently, are exposed to external and global changes in trade and markets. Many countries have traditionally had monoculture economies, relying on preferential trade agreements for their main exports. Some governments have tried to reduce dependence on monoculture agriculture by promoting agricultural diversification. However, in some countries, such as St.

Lucia, there is concern about high rates of deforestation of natural forests in response to the push for diversification.

Trade barriers for Caribbean exports to North America and Europe have increased in recent years, and the region's export markets have been threatened by major trade agreements, such as the North American Free Trade Association (NAFTA) and the Economic Partnership Agreement (EPA), and preferential markets for products such as bananas and rum have been lost.

Table 6.2: Environmental vulnerability index ratings of selected SIDS.

Extremely Vulnerable	Highly Vulnerable	Vulnerable	At risk	Resilient
Barbados Virgin Islands British Guadeloupe Jamaica St. Lucia Virgin Islands of United States	Cayman Islands Cuba Dominican Republic Grenada Haiti Martinique Montserrat * Netherlands Antilles Puerto Rico San Cristóbal and Nieves St. Vincent and the Grenadines	Eel Antigua and Barbuda Aruba * Aruba * Aruba * Aruba * Aruba * Aruba * Aruba * Aruba * Aruba Turks and Caicos * Caicos * Caicos * Caicos	The Bahamas	None

Many countries also have high levels of external debt but small taxable populations, posing a challenge to their long-term economic viability. Their openness to foreign trade also increases their susceptibility to IAS.

Preliminary rankings on the SIDS environmental vulnerability index, which measures ecological fragility and economic vulnerability, show that 17 of the countries/territories can be classified as extremely vulnerable or highly vulnerable, four as vulnerable and one as at risk, while none are assessed as resilient (Table 6.2).

Global climate change

Climate change is expected to become a major driver of environmental change in the hotspot and is already causing substantial impacts, none of which are fully understood (Section 10.3).

4.2.2 Barriers to wildlife conservation

There are several constraints that must be overcome to address the environmental threats described above and to achieve more effective conservation of biodiversity and ecosystem services. This section reviews the main threats identified in the national ecosystem profile consultations, which CEPF reinvestment will seek to address.

Poor land use planning

Because many environmental problems and risks arise from or are exacerbated by the pattern of human land use, the quality of urban and rural planning is often critical to achieving environmental sustainability.

In the small West Indies, with its dense coastal populations, inappropriate land use can have much more significant impacts on the environment than in the larger states, and there is less room for error in land-use planning and management (Griffith and Ashe, 1993).

Land use planning for agriculture, tourism, industry, forestry and urban development is still largely limited to individual sectors, with little consideration of the impacts of these plans on other economic sectors or the environment.

Strategic environmental assessments, for example, are not routinely conducted in the Caribbean, and the environmental costs of development are generally not incorporated into national accounts.

Furthermore, although the location of many sites important for biodiversity conservation and the provision of ecosystem services has been identified through surveys and mapping exercises in recent years, this information is not yet fully integrated into decision-making in planning processes. Consequently, ecologically important sites are still subject to inappropriate development.

Constraints in capacity and financial resources for biodiversity conservation and environmental management

While Antillean governments have made significant efforts to strengthen their institutional and individual capacity (in terms of personnel and financial resources) in the areas of biodiversity conservation, waste management, integrated watershed management, and climate change and disaster mitigation, the lack of adequate capacity remains a major barrier to effective environmental management and sustainable development.

Most island states have populations of less than one million people (Section 7.1.2), with small skilled labor pools and very limited government budgets for the environmental sector, severely limiting capacity building efforts. This has a particular impact on staffing in government entities.

Individuals often seek higher education outside the region due to limited training opportunities in natural resource management and biodiversity conservation at Caribbean universities, often traveling to the United States or Canada, where many choose to stay because of better salaries and career opportunities. If they return, many enter the private sector or seek employment in unrelated but higher paying professions in the financial or legal sectors. Consequently, the region's "brain drain" and difficulties with staff retention by government entities continue to be major issues affecting capacity in many Caribbean countries. Even in the larger islands, the size of government environmental departments, in terms of staff and financial resources allocated, is not sufficient.

The need for capacity building has been highlighted in many of the NBSAPs, national environmental action plans, national protected area gap analyses and other national strategies and plans.

Protected area management also stands out as generally weak in the Caribbean. Despite considerable investment in recent years by governments and external donors, protected area management agencies are still under-resourced and many protected areas have minimal or no management and are threatened by encroachment and illegal activities.

Many donor-funded biodiversity conservation projects also have significant evaluation capacity and building elements, but due to the lack of trained personnel in the target countries, external consultants are still often used in project implementation, which does not address the long-term problem (Renard and Geoghegan, 2005).

There is a lack of qualified field biologists in government agencies to carry out the survey and assessment work needed to generate terrestrial (and marine) natural resource management regulations. As a result, field surveys are often conducted through the NGO community or universities.

Perhaps the biggest capacity problem is the lack of personnel and resources among the entities responsible for monitoring, surveillance and enforcement of current national legislation and regulations governing biodiversity conservation and environmental management (such as monitoring and enforcement of compliance with EIAs and planning restrictions). This is a particular concern given the continuing pressures of tourism, urban and industrial development in the region. In several countries, policies and legislation are largely considered adequate, but lack of enforcement and monitoring, as well as lack of coordination between entities, undermine their implementation.

Capacity problems often boil down to a lack of financial resources. Some funds have been short-term and project-driven and have rarely been strategic. This situation works against sustainable environmental management, whether in civil society, the private sector or the government sector.

Lack of knowledge and understanding of the importance of biodiversity and ecosystem services.

In addition to lack of knowledge, there is a lack of awareness and limited understanding of the ecological, economic, social and cultural values of biodiversity, the costs of its loss and its critical importance to human health and well-being by decision makers (ministers, politicians, policy advisors, economists and land use planners) and the general public in the Caribbean. Even in relatively developed countries, such as Puerto Rico, the level of public awareness of national biodiversity is low.

Some governments are using a longer-term strategy, with an emphasis on improving the coverage of environmental issues in the national school curriculum. Barbados, for example, has introduced environmental and development issues into teacher training programs, while environmental education is an integral part of primary and secondary school curricula in The Bahamas. These initiatives will ultimately increase the proportion of the population with environmental awareness and interest, leading to a greater call for politicians and other decision-makers to adequately address environmental issues, and an overall increase in people with technical skills for biodiversity conservation.

There is also a relatively low level of exchange of data and information and lessons learned on environmental issues between countries.

Vested interests, corruption and lack of political will

There have been several important regional environmental agreements in the Caribbean in the past, including the Georgetown Agreement in 1975, the Nassau Understanding in 1984, the CARICOM Ministerial Conference on the Environment in 1989, the St. George's Declaration in 2000, the OECS Environmental Management Strategy in 2001 and the OECS Development Charter in 2002. However, commitment among high-level decision-makers has not yet translated into the necessary political support for biodiversity conservation.

Short-term, and often shifting, national economic and political interests often take precedence over long-term social and environmental impacts. This lack of political will is evidenced by continued permission for destructive developments in ecologically sensitive areas, often as a result of strong lobbying by vested economic interests who argue that the costs of environmental protection and safeguards reduce international competitiveness.

Weak and inefficient policies and legislation

Although there have been improvements in national policy frameworks and progress has been made in updating and harmonizing environmental policies and laws in many countries in the region in recent years, there are still policy gaps and challenges. For example, legislation for the establishment of private reserves and co-management of protected areas is lacking in most Caribbean countries.

The Dominican Republic is a notable exception in this regard, having developed legislation for both. In addition, while some countries, such as Barbados, the French overseas regions and the U.S. territories have legislation dealing specifically with the coastal zone, many countries do not have special instruments to regulate development in this ecologically critical area. In addition, there has been limited integration of biodiversity conservation and sustainable environmental management objectives into non-environmental sector policies and legislation, and relatively little coverage in development plans and sectoral plans. For example, regulations on the use of pesticides and fertilizers tend to be very weak or non-existent; as a result, these materials are often applied in excessive amounts, which do not improve productivity but instead contaminate ground and surface water supplies and have a negative impact on wildlife.

In some cases, laws lack clear regulations to provide guidance to developers, which is compounded by inadequate environmental codes and standards for land development, construction, resource use and waste management, limiting the ability of authorities to enforce environmental protection. In addition, individual developments are often carried out without adequate assessment of their impact on the local environment and natural resources.

Lack of efficiency in institutional frameworks, networks and collaboration

Previous assessments have also identified several deficiencies in institutional frameworks and operations that constrain the effectiveness of environmental management. Chief among these is that management authority for environmental issues (biodiversity, forests, water reservoirs, protected areas, etc.) is often divided among various ministries and other statutory bodies, whose responsibilities overlap or are not well defined, resulting in inefficiency, lack of accountability, or inaction. This is compounded by the lack of institutional mechanisms for coordination and collaboration among the numerous actors and programs.

The lack of coordination and collaboration between governments and CSOs is paralleled by insufficient donor coordination at the country level. This has been identified as a major problem in Haiti (Smucker *et al.*, 2007), where greater collaboration among donors at the policy level is considered a critical need, as is the targeting of field interventions. Another critical point is the need to share documents of project strategies, successes and activities among different donors.

The prevailing view of the environment as a niche problem is reflected in the lack of systematic integration of environmental objectives into overall sectoral policies and programs, which is partly a reflection of the poor understanding among decision-makers of the linkages between biodiversity and ecosystem services and local livelihoods, employment and national economies. This results in politically weak and underfunded environmental entities, and biodiversity conservation policies continue to be perceived as incompatible with and constraining economic development.

Despite the presence of national sustainable development strategies in many countries that highlight the importance of biodiversity, adequate solutions are not apparent. However, attitudes towards the environment at higher levels appear to be changing, due to a greater awareness of the impacts of climate change, which is having a real effect in the insular Caribbean.

Inadequate public participation in decision-making processes
National and local governance frameworks for environmental planning and management vary widely among countries, but governments are generally highly centralized, often with high levels of state control.

More recent national policy frameworks include provisions for private sector and public stakeholder participation in environmental and development decision-making, such as through National Sustainable Development Councils. Stakeholder participation is also promoted in many regional and international initiatives involving Caribbean governments.

Caribbean CSOs have been becoming more involved in national and regional policy and decision-making processes and are increasingly recognized as important actors in these spheres. However, while CSO participation in governance processes is growing, it is not always adequately supported by mechanisms that facilitate meaningful participation or consider the conditions under which CSOs operate.

Limited technical and scientific knowledge and scarce availability of information

While the island Caribbean countries have shown significant improvement in research and assessment of their living natural resources in recent years, there are considerable gaps in baseline data and a lack of accurate and up-to-date information, which limits effective evidence-based decision making for biodiversity conservation, land use planning, EIAs and environmental monitoring. Lack of data and information also complicates regulatory compliance. Information is often scattered and difficult to access, with poor coordination and linkages between databases.

Attempts have been made to overcome some of these problems through the creation of national information exchange mechanisms, while the European Union-funded *Biodiversity Information for Development (BID)* project is currently attempting to strengthen national biodiversity data and information facilities.

One area of particular importance is the need to demonstrate the economic and social benefits of ecosystem services in the Caribbean to policy makers, planners and other decision makers.

To date, there have been few studies on this topic in the region. In addition, little attention has been paid to the economic costs of the loss of ecosystem services, which can be enormous and lead directly to loss of life, for example, due to reduced protection against extreme weather events. In addition, the lack of adequate accounting for the value of ecosystem services means that the costs of environmental protection tend to be overestimated and the benefits underestimated. There is also a general need for greater collection, analysis and translation of technical information into formats suitable for decision-makers and the public to better communicate issues such as the value of biodiversity for tourism, the impacts of development on biodiversity and the costs of biodiversity loss.

5. FAUNA OF TRINIDAD &TOBAGO, A SPECIAL CASE IN THE ANTILLES

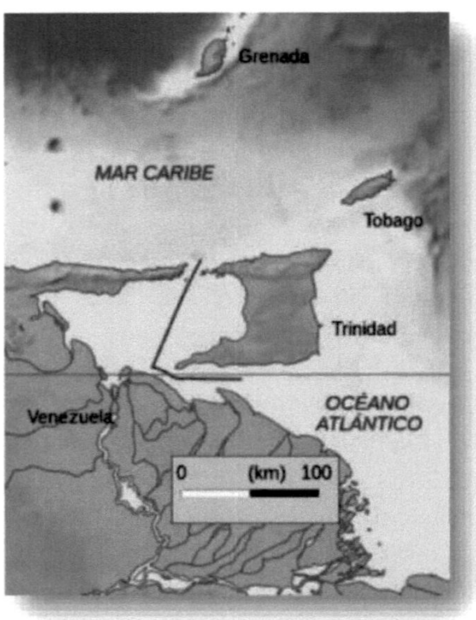

Because Trinidad and Tobago lies on the continental shelf of South America, and was once physically connected to the South American continent, its biological diversity is different from that of most of the other West Indies, and has much more in common with that of Venezuela.

The main ecosystems are: coastal and marine (coral reefs, mangroves, open ocean and seagrass); forest; freshwater (rivers and streams); karst; artificial ecosystems (agricultural land, freshwater dams, secondary forests); and savanna.

On August 1, 1996, Trinidad and Tobago ratified the 1992 Rio Convention on Biological Diversity and has developed a biodiversity action plan and four reports describing the country's contribution to biodiversity conservation. These reports formally recognize the importance of biodiversity to the well-being of the country's population through the provision of ecosystem services.

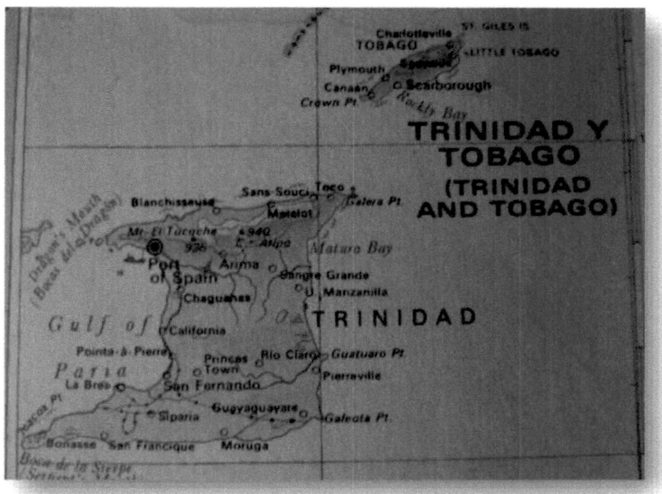

Vertebrate information is good, with 472 bird species (2 endemic), about 100 mammal species, about 90 reptile species (some endemic), about 30 amphibian species (including several endemics), 50 freshwater fish and at least 950 marine fish.

Mammal species include the ocelot, West Indian manatee, collared peccary (known locally as quenk), red-rumped agouti, lapwing, red deer, neotropical otter, weeping capuchin and red howler monkey.

Among the larger reptiles are five species of marine turtles that nest on the beaches, the green anaconda, the boa constrictor and the spectacled caiman. There are at least 47 species of snakes, of which only four are dangerous venomous species (only in Trinidad and not in Tobago), lizards such as the green iguana, the *Tupinambis cryptus and* some species of freshwater and terrestrial turtles. are present.

Of the amphibians, the golden tree frog is found only on the highest peaks of the Northern Cordillera of Trinidad and, nearby, on those of the Paria Peninsula in Venezuela. Marine life is abundant, with several species of sea urchins, corals, lobsters, anemones, starfish, manta rays, dolphins, porpoises and whale sharks present in the islands' waters.

The introduced pterois is considered a pest, as it eats many native fish species and has no natural predators; efforts are currently underway to reduce the number of this species.

The country contains five terrestrial ecoregions: the humid forests of Trinidad and Tobago, the dry forests of the Lesser Antilles, the dry forests of Trinidad and Tobago, the xeric scrublands of the Windward Islands and the mangroves of Trinidad.

Trinidad and Tobago is especially noted for its large number of bird species, and is a popular destination for birdwatchers. Notable species include the scarlet ibis, cocrico, egret, glossy cowbird, bananaquit, oilbird and several species of honeycreeper, trogon, toucan, parrot, tanager, tanager, woodpecker, antbird, kites, hawks, boobies, pelicans and vultures; there are also 17 species of hummingbirds, including the tufted coquette, which is the third smallest in the world.

Information on invertebrates is scattered and very incomplete. Some 650 butterflies, at least 672 beetles (from Tobago alone) and 40 corals have been recorded. Other notable invertebrates include the cockroach, leaf-cutter ant and numerous species of mosquitoes, termites, spiders and tarantulas.

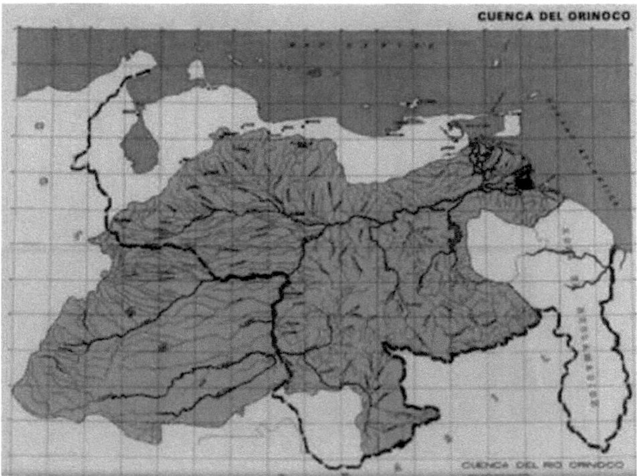

Although the list is far from complete, 1 647 species of fungi, including lichens, have been recorded. It is likely that the true total number of fungi is much higher, since the generally accepted estimate is that only 7% of all fungi in the world have been discovered so far. In a first effort to estimate the number of endemic fungi, 407 species were tentatively listed.

Despite significant logging, forests still cover about 40% of the country, and there are about 350 different species of trees. One notable tree is the manchineel, which is extremely poisonous to humans, and even just touching its sap can cause severe blistering of the skin; the tree is often covered with warning signs. The country obtained an average Forest Landscape Integrity Index score of 6.62/10 in 2019, ranking 69th globally out of 172 countries.

Threats to the country's biodiversity include excessive hunting and poaching, habitat loss and fragmentation (mainly due to forest fires and land clearing for quarrying, agriculture, illegal occupation, real estate and industrial development and road construction), water and air pollution, and the introduction of exotic species.

BIBLIOGRAPHIC REFERENCES

2 The Final Logframe Report for the West Indies Biodiversity Hotspot (2010 - 2016) is available for download at: https://www.cepf.net/sites/default/files/caribbean- islands-final-logframe-report-2016-english.pdf.

3 CANARI Policy Brief No. 22 "Effectively Supporting Caribbean Civil Society for Biodiversity Conservation and Rural Development: results and recommendations from the Critical Ecosystem Partnership Fund 2010 - 2016" is available for download at https://canari.org/effective-support-for-caribbean-civil-society-for-biodiversity-conservation-and- rural-development-results-and-recommendations-from-the-critical-ecosystem-partnership-fund- 2010-2016/

CANARI Policy Brief No. 23 "Effective Donations to Caribbean Civil Society: Lessons and Innovation from CANARI's Experience as an Intermediary Organization" can be downloaded at: https://canari.org/civil-society-and-governance/effective-grant-making-to- caribbean-civil-society-lessons-and-innovation-from-canaris-experience-as-an-intermediary-organization/

4 Caribbean Coastal Area Management Foundation, Jamaica Environment Trust, Caribbean Wildlife Alliance, Conservation Strategy Fund, International Iguana Foundation, Birds Caribbean, World Resources Institute and the Windsor Research Center.

BIBLIOGRAPHIC REFERENCES

Abeyasuriya, K., Nugapola, N., Perera, M., Karunaratne, W., and Karunaratne, S. (2017) Effect of dengue mosquito control insecticide thermal fogging on non-target insects. *International Journal of Tropical Insect Science.* 37(1), 11-18.

Acevedo-Rodriguez, P., and Strong, M. (2007) *Flora of the West Indies.* Washington, D.C.: National Museum of Natural History and The Smithsonian Institution.

Acevedo-Rodríguez, P., and Strong, M. (2008) Floristic richness and affinities in the West Indies. *Botanical Review,* 74(1), 5-36.

Agarwal, A., Cashore, B, Hardin, R., Shepherd, G., Benson, C., and Miller, D. (2013) *Economic contributions of forests. Background Paper 1.* United Nations Forum on Forests. Retrieved from http://www.un.org/esa/forests/wp-content/uploads/2015/12/EcoContrForests.pdf.

Aho, E. (2017) *Shrinking space for civil society: Challenges in implementing the 2030 agenda.* Stockholm: Forum Syd. Retrieved from http://www.forumsyd.org/PageFiles/8150/PO150943_

AIMS (2002) *Status of coral reefs of the world: 2002* (C. Wilkinson, Ed.) Townsville, Queensland, Australia: Australian Institute of Marine Sciences.

ALBA (2017) *Statement by the Bolivarian Alliance of the Peoples of Our Americas (ALBA) at the opening ceremony of COP 23 of the UNFCCC.* World Conference Centre, Bonn, 6 November 2017.

Alleyne, T. C. (2017) *Unleashing growth and strengthening resilience in the Caribbean.* Washington, D.C.: International Monetary Fund.

Anderson, E. C. (2008) *Potential impacts of climate change on biodiversity in Central America, Mexico, and the Dominican Republic.* Panama: CATHALAC/USAID.

Andrews, S., Cumberbatch, J., and Hinds, C. (2012) *Advancing sustainable tourism. A regional sustainable tourism situational analysis: Caribbean* . UNEP and Global Partnership for Sustainable Tourism.

Anglo-American Caribbean Commission (AACC) (1945) *Guide to commercial shark fishing in the Caribbean area.* Washington, D.C: U.S. Fish and Wildlife Service, Anglo-American Caribbean Commission.

Arias, Y (2018, February 9, 2018) KBAs and corridors in La Selle-Bahoruco- Jaragua Biosphere Reserve. *Telephone interview with David Díaz.*

Arkema, K., Fisher, D., and Wyatt, K. (2017) *Economic valuation of ecosystem services in Bahamian marine protected areas.* Prepared for BREEF by The Natural Capital Project, Stanford University.

Association of Caribbean States (ACS) (2017) ACS. [Web page]. Retrieved from http://www.acs-aec.org/

Associated Press. (2017, January 10) Puerto Rico governor pursues freedom of information law. *News 1130.com.* Retrieved from http://www.news1130.com/2017/01/10/puerto-rico-governor-pursues- freedom-of-information-law/.

Atteridge, A., Canales, N. and Savvidou, G. (2017) *Climate finance in the Caribbean region's Small Island Developing States.* Working Paper No. 2018-08. Stockholm: Stockholm Environment Institute - Stockholm Centre.

Attzs, M., Maharaj, M., Boodhan, G. (2014) *Survey and assessment of environmental taxes in the Caribbean.* Policy Brief No. IDB-PB-188. Washington, D.C.: Inter-American 278.

Development Bank. Retrieved from http://www20.iadb.org/intal/catalogo/PE/2014/14506.pdf

AWID. (2017) *Caribbean unity in support of Tambourine Army activist Latoya Nugent.* [Web page]. Retrieved from https://www.awid.org/news-and- analysis/caribbean-unity-support-tambourine-army-activist-latoya-nugent on August 15, 2017.

Bahamas Environment, Science and Technology Commission (2001) *First national communication.* Nassau: Commonwealth of The Bahamas.

Bahamas National Trust (2017) Corporate engagement launch. *Trust Notes* 12(31)

Baldwin, C., and Robertson, R. (2014) A new Liopropoma sea bass (Serranidae, Epinephelinae, Liopropomini) from deep reefs off Curaçao, southern Caribbean, with comments on depth distributions of western Atlantic liopropomins. *ZooKeys,* 409, 71-92. doi:
10.3897/zookeys.409.7249

Baldwin, C., and Robertson, D. (2015) A new, mesophotic Coryphopterus goby (Teleostei, Gobiidae) from the southern Caribbean, with comments on relationships and depth distributions within the genus. *ZooKeys,* 513, 23-142. doi:10.3897/zookeys.513.9998.

Baldwin, C., Pitassy, D., and Robertson, D. (2016a) A new deep-reef scorpionfish (Teleostei, Scorpaenidae, Scorpaenodes) from the southern Caribbean with comments on depth distributions and relationships of western Atlantic members of the genus. *ZooKeys,* 606, 141-158. doi:10.3897/zookeys.606.8590

Baldwin, C., Robertson, R., Nonaka, A., and Tornabene, L. (2016b) Two new deep-reef basslets (Teleostei Grammatidae, Lipogramma), with comments on the eco-evolutionary relationships of the genus. *ZooKeys,* 638, 45-82. doi:10.3897/zookeys.638.10455.

BirdLife International. (2017) *Handbook of the birds of the world and BirdLife International digital checklist of the birds of the world.* Version 9.1. Cambridge, UK. Retrieved from.
http://datazone.birdlife.org/userfiles/file/Species/Taxonomy/Bird Life_Che cklist_Version_91

BirdLife International and Grupo Jaragua (2015) *Ecosystem services of Sierra de Bahoruco National Park, Dominican Republic.* Santo Domingo: BirdLife International and Grupo Jaragua.

BirdsCaribbean (2017, September 11) *After the storm.* Retrieved from http://www.birdscaribbean.org/2017/09/after-the-storm/ on October 15, 2017.

Blackman, D., Epanchin-Niell, R., Siikamki, J. and Velez Lopez, D. (2015) *Biodiversity conservation in Latin America and the*

Caribbean: Prioritizing policies. New York: Routledge

Belpomme D. (2007) *Rapport d'expertise et d'audit externe concernant la pollution par les pesticides en Martinique. Conséquences agrobiologiques, alimentaires et sanitaires et proposition d'un plan de sauvegarde en cinq points.* Paris: Association pour la Recherche Thérapeutique Anti-Cancéreuse http://www.nord-nature.org/environnement/energie/com/rapport-martinique%20chlordecone.pdf

Berger, L., Speare, R., Daszak, P., Green, D.E., Cunningham, A.A., Goggin, C.L., Slocombe, R., Ragan, M.A., Hyatt, A.D., McDonald, K.R., Hines, H.B., Lips, K.R., Marantelli, G., Parkes, H. (1998) Chytridiomycosis causes amphibian mortality associated with population declines in the rain forests of Australia and Central America. *PNAS* 95: 9031-9036. 279

Bosch, J., and Rincon, P. A. (2008) Chytridiomycosis-mediated expansion of *Bufo bufo* in a montane area of Central Spain: An indirect effect of the disease. *Diversity and Distributions* 14: 637-643.

Bovarnick, A., Alpizar, F., and Schnell, C. (2010) *The importance of biodiversity and ecosystems in economic growth and equity in Latin America and the Caribbean: An economic valuation of ecosystems.* New York: United Nations Development Programme.

Bowen, G. A. (2015) Caribbean civil society: Development role and policy implications. *Nonprofit Policy Forum,* 4(1), 81-97.

Brautigam, A. and Eckert, K. L. (2006) *Turning the tide: Exploitation, trade and management of marine turtles in the Lesser Antilles, Central America, Colombia and Venezuela.* Cambridge, UK: TRAFFIC International.

Breuil, M. (2013) Morphological characterization of the common iguana Iguana iguana (Linnaeus, 1758), of the Lesser Antillean Iguana Iguana delicatissima Laurenti, 1768 and of their hybrids. *Bull. Soc. Herp., 147,* 309-346.

Brooks, T. M., Mittermeier, R. A., Mittermeier, C. G., Da Fonesca G. A. B., Rylands, A. B., Konstant, W. R., Flick, P., Pilgrim, J.,

Oldfield, S., Magin, G., and Hilton-Taylor, C. (2002) Habitat loss and extinction in the hot spot of biodiversity. *Conservation Biology* 16:909-923.

Brown, D. (2017, April 26) Government cuts $700,000 subsidy to St. Lucia National Trust. *Caribbean News Service*. Retrieved from https://caribbeannewsservice.com/now/government-cuts-700000- subvention-to-st-lucia-national-trust/ on September 15, 2017.

Brown, N. A., and Bennett, N., G (2010) *Consolidating change: lessons from a decade of experience in mainstreaming Local forest Management in Jamaica. Technical Report No. 390.* Laventille: Caribbean Natural Resources Institute (CANARI).

Brown, N.A., Geoghegan T., and Renard, Y. (2007) *A situation analysis for the wider Caribbean.* Gland: IUCN.

Brown, N. A.; Diaz, D.; Angarita, I.; Cadiz, H. N.; Boodram, G. C.; Banting, A.; Haeded, F. (2019) *Hospot of Caribbean Island Biodiversity. Ecosystem Profile.* CANARI - BirdLife International - IUCN - NYBG. New York, 533 pp.

Brundige, E; Dominguez Cisneros, L., Peñalver, E. M.; and Spitz, L. (2017) U.S. Nonprofit activity in Cuba: The Cuban context. *Cornell International Law Journal:* Vol. 50: No. 2, Article 2. Retrieved from.
https://scholarship.law.cornell.edu/cilj/vol50/iss2/2

Burke, L., and Maidens, J. (2004) *Reefs at risk in the Caribbean.* Washington D.C.: World Resources Institute.

Burrowes, P. A, Joglar, R. L, Green, D. E. (2004) Potential causes for amphibian declines in Puerto Rico. *Herpetologica* 60: 141-154

Cano-Ortiz, A., Musarella, C., Piñar Fuentes, J., and Cano, E. (2016) Distribution patterns of endemic flora to define hotspot on Hispaniola. *Systematics and Biodiversity,* 14(3), 261-275. doi:10.1080/14772000.2015.1135195.

Cap-Net (2015) *Integrated Water Resources Management as a Tool for Adaptation to Climate Change with Caribbean Case Studies.* Cap-Net.

Caribbean Agricultural Research Institute (CARDI) (n.d.) *Climate Change and Water Availability in the Caribbean. Policy Brief.* St.

Augustine: Caribbean Agricultural Research Institute. Retrieved from http://www.cardi.org/wp-content/uploads/2012/02/POLICY-BRIEF- DRAFT_CC-and-Water-Availability.pdf

Caribbean Biodiversity Fund (2014) Caribbean Biodiversity Fund factsheet. Nassau: Caribbean Biodiversity Fund. Retrieved from 280

http://www.caribbeanchallengeinitiative.org/images/articles/CBF _Fact_S heet_-_May_5_2014-1.pdf.

Caribbean Coastal Area Management Foundation and Jamaica Environment Trust (2013) *The Goat Islands/Portland Bight Protected Area. The proposed site for a transshipment port in Jamaica.* Kingston: Caribbean Coastal Area Management Foundation and Jamaica Environment Trust.

Caribbean Community Climate Change Centre (CCCCCCC) (2009) Climate change and the Caribbean: A regional framework for achieving development resilient to climate change (2009-2015) Belmopan: Caribbean Community Climate Change Centre.

Caribbean Development Bank (2014a) *Caribbean economic review and outlook for 2015.* St. Michael, Barbados: Caribbean Development Bank.

Caribbean Development Bank (2014b) *A new paradigm for Caribbean development: transitioning to a green economy*. St. Michael, Barbados: Caribbean Development Bank.

Caribbean Development Bank (2016) *2016 economic review/2017 forecast.* Bridgetown: Caribbean Development Bank.

Caribbean Disaster Emergency Management Agency (CDEMA) (2017a*) Hurricane Maria situation report #5.* Bridgetown: Caribbean Disaster Emergency Management Agency.

Caribbean Disaster Emergency Management Agency (CDEMA) (2017b) *Hurricane Irma situation report #9.* Bridgetown: Caribbean Disaster Emergency Management Agency.

Caribbean Environment Programme (CEP) (2003) *The Caribbean Environment Programme: Promoting regional co-operation to protect the marine environment.* Kingston: UNEP-CEP

Caribbean Environmental Health Institute (2012) *Programme on improving management of coastal resources and the conservation of marine biodiversity in selected CARICOM Countries: Baseline study*. Castries: CEHI.

Caribbean Institute for Meteorology and Hydrology (CIMH) (2016) *Caribbean climatology - Caribbean Regional Climate Centre.* [Web page]. https://rcc.cimh.edu.bb/caribbean-climatology/

Caribbean Natural Resources Institute (CANARI) (2005) *Governance and civil society partnership in sustainable development in the Caribbean.* CANARI Policy Brief No 7. Laventille: CANARI

Caribbean Natural Resources Institute (CANARI) (2012) *Community forestry in the Caribbean: A regional synthesis.* Laventille: CANARI.

Caribbean Natural Resources Institute (CANARI) (2013) *Summary report for the regional workshop Critical Ecosystem Partnership Fund (CEPF) Caribbean Islands Biodiversity Punto caliente investment (2010-2015) mid-term evaluation.* Laventille: CANARI.

Caribbean Natural Resources Institute (CANARI) (2017a) *OECS green economy diagnostic: Exploring opportunities for green economy transformation in the Eastern Caribbean.* Laventille: CANARI.

Caribbean Natural Resources Institute (CANARI) (2017b) *Effective organisational capacity building of civil society organisations: Lessons from the Climate ACTT Project.* CANARI Policy Brief No. 24. Laventille: CANARI.

Caribbean Natural Resources Institute (CANARI) (2017c) *Implementing climate change action: A toolkit for Caribbean civil society organisations* . Laventille: CANARI. 281

Caribbean Regional Fisheries Mechanism Secretariat (CRFM) (2015) *CRFM statistics and information report - 2014.* Kingstown: Caribbean Regional Fisheries Mechanism Secretariat.

Caribbean Regional Fisheries Mechanism (2017) *CRFM.* [Web page]. Retrieved from http://www.crfm.int/

Caribbean Tourism Organisation (2015) *Latest statistics 2015.* Warrens, St. Michael, Barbados: CTO. Retrieved from. http://www.onecaribbean.org/wp-content/uploads/Lattab15_FINAL.pdf.

CARICOM (2014) *Strategic plan for the Caribbean Community 2015 - 2019: Repositioning CARICOM. Vol 2 - The Strategic Plan.* Georgetown: Caribbean Community Secretariat.

CARICOM. (2016) *COP 22- CARICOM close out brief V3 (2)* Georgetown: Caribbean Community Secretariat.

CARICOM. (2017) *Caribbean Centre for Renewable Energy and Energy Efficiency (CCREEE)* Georgetown: Caribbean Community Secretariat. Retrieved from.

CARICOM (r.d.) *Capacity building related to the implementation of Multilateral Environmental Agreements (MEAS) in African, Caribbean and Pacific (ACP) Countries - The Caribbean Hub.* [Web page]. Retrieved from https://caricom.org/projects/detail/capacity-building- related-to-the-implementation-of-multilateral-environment

Cashman, A., Nurse, L. and John, C. (2010) Climate change in the Caribbean: The water management implications. *The Journal of Environment and Development,* 19(1), 42-67.

Caujapé-Castells, J. (2011) Preface. In D. Bramwell, and J. Caujapé-Castells (Eds.), *The biology of island floras* (pp. xiii-xvi) Cambridge: Cambridge University Press.

CEPF (2010) *Ecosystem profile: the Caribbean Islands Biodiversity Hot Spot.* Arlington, VA : Critical Ecosystem Partnership Fund.

Cesar, H. S., Ohman, M. C., Espeut, P., Honkanen, M. (2000) An economic valuation of Portland Bight, Jamaica: An integrated terrestrial and marine protected area. In Cesar, H. S. (Ed.), *Collected essays on the economics of coral reefs* (pp. 203-214) Borâs, Sweden: CORDIO.

Chao, S. (2013) *Economic impact of non-communicable disease in the Caribbean.* Washington, D.C.: The World Bank Group.

Cheng, T. L., Rovito, S. M., Wake, D. B., and Vredenburg, V. T. (2011) Coincident mass extirpation of neotropical amphibians with the emergence ofthe infectious fungal pathogen *Batrachochytrium dendrobatidis. Proceedings of the National Academy of Sciences of the United States of America,* 108(23), 9502-9507. Retrieved from http://doi.org/10.1073/pnas.1105538108

CIVICUS (2017a) *Civic space in the Americas.* Johannesburg:

CIVICUS.

CIVICUS (2017b) *People power under attack: Findings from the CIVICUS Monitor.* Johannesburg: CIVICUS.

Commission for Environmental Cooperation for North America (CEC) (2001) *Preventing the introduction and spread of aquatic invasive species in North America, 28-30 March 2001.* Montreal: CEC.

Congressional Task Force on Economic Growth in Puerto Rico (2016) *Report to the House and Senate 114th Congress.* Washington, D.C.: United States Congress. 282

Connor, D. (2017, April 13) Racer snakes may face development threat. *Discover Wildlife.* Retrieved from. http://www.discoverwildlife.com/news/racer-snakes-may-face-development-threat

Consorcio Ambiental Dominicano (2015) *CEPF final project completion report: Sustainable financing and establishment of private reserves for biodiversity conservation in Loma Quita Espuela and Loma Guaconejo, Dominican Republic.*

Curley, R. (2017) *Trip savvy: Green globe certified hotels and attractions in the Caribbean.* Retrieved from https://www.tripsavvy.com/green-globe-   certified-hotels-and-attractions-in-the-caribbean-1487670 on November 18, 2017.

Daltry, J. (2009) *Biodiversity assessment of Saint Lucia's forests, with management recommendations.* Technical Report No. 10 to the National Forest Demarcation and Bio-Physical Resource Inventory. Helsinki: FCG International Ltd.

Daltry, J. (2018, January 19) Globally threatened reptiles updates in the Caribbean. *E-mail correspondence with David Díaz.*

Dart, T. (2017, March 31) Caribbean resort project draws heat over threat to vulnerable species. *The Guardian.* Retrieved from. https://www.theguardian.com/world/2017/mar/31/st-lucia-pearl-of-the- caribbean-resort-environmental-threat on March 3, 2018.

Daszak, P., Cunningham, A. A., and Hyatt, A. D. (2000) Emerging infectious diseases of wildlife-threats to biodiversity and human health. *Science* 287: 443-449.

Debrot, A.O. and Bugter, R. (2010) *Climate change effects on the biodiversity of the BES islands; Assessment of the possible* consequences for the marine and terrestrial ecosystems of the Dutch Antilles and the options for adaptation measures. Alterra-report 2081; IMARES-report C118/10. Wageningen: Alterra.

Díaz, L.M., Cádiz, A., Agustín, A., Chong, A., and Silva, A. (2007) First report of chytridiomycosis in a dying toad *(Anura: Bufonidae)* from Cuba: A New Conservation Challenge for the Island. *EcoHealth* 4: 172.

Dolisca, F. (2005) *Population pressure, land tenure, deforestation and farming systems in Haiti: The case of Forêt des Pins Reserve* (Doctoral dissertation) Auburn, Alabama: Auburn University.

Dudley, N., Buyck, C., Furuta, N., Pedrot, C., Renaud, F., and Sudmeier- Rieux, K. (2015) *Protected areas as tools for disaster risk reduction. A handbook for practitioners.* Tokyo and Gland, Switzerland: MOEJ and IUCN.

Dudley, N., Stolton, S., Belokurov, A., Krueger, L., Lopoukhine, N., MacKinnon, K., Sandwith, T., and Sekhran, N. (2010) *Natural solutions: Protected areas helping people cope with climate change.* Gland: IUCNWCPA, TNC, UNDP, WCS, The World Bank, WWF.

Eckstein, D., Künzel, V., and Schafer, L. (2017) *Global climate risk index 2018.* Bonn: Germanwatch e.V.

Economic Commission for Latin American and the Caribbean (ECLAC) (2006) *Changing population age structures and their implications on socio-economic development in the Caribbean.* Santiago, Chile: ECLAC. (LC/CAR/L.98) https://www.cepal.org/publicaciones/xml/3/27113/L.98.pdf Economic Commission for Latin American and the Caribbean (ECLAC) (2008) *Impact of changes in European Union import regimes for sugar, banana and rice on selected CARICOM countries.* Santiago, Chile: ECLAC. (LC/CAR/L/168)

Economic Commission for Latin American and the Caribbean (ECLAC) (2009) *Statistical yearbook for Latin America and the Caribbean 2008.* Santiago, Chile: ECLAC. 283 Retrieved from http://repositorio.cepal.org/bitstream/handle/11362/923/1/S2008 691_mu. pdf

Economic Commission for Latin American and the Caribbean

(ECLAC) (2011) *The economics of climate change in the Caribbean.* Caribbean Development Report III. Santiago, Chile: ECLAC.

Economic Commission for Latin American and the Caribbean (ECLAC)

(2015) *Latin America and the Caribbean: Looking ahead after the Millennium Development Goals.* Regional monitoring report on the Millennium Development Goals in Latin America and the Caribbean. Santiago, Chile: ECLAC. (LC/G.2646) Retrieved from http://repositorio.cepal.org/bitstream/handle/11362/38924/S150 0708_en. pdf

Economic Commission for Latin American and the Caribbean (ECLAC)

(2016) *Foreign direct investment in Latin America and the Caribbean* . Santiago, Chile: ECLAC

Economic Commission for Latin American and the Caribbean (ECLAC)

(2017) *Economic activity in Latin America and the Caribbean will expand 1.2% in 2017 and 2.2% in 2018.* [Press release]. Retrieved from https://www.cepal.org/en/pressreleases/economic-activity-latin-america- and-caribbean-will-expand-12-2017-and-22-2018

Edwards, P. (2011) *Ecosystem service valuation of Cockpit Country .* Sherwood Content: Windsor Research Centre.

Edwards, P. (2013) *Ecosystem services of the Coral Spring and Mountain Spring Protected Area, Jamaica.* Sherwood Content: Windsor Research Centre.

El Dia. (2017, January, 24) Dominican Republic will have new regulation for NGOs. *El Dia.* Retrieved from http://eldia.com.do/republica- dominicana-tendra-nueva-regulacion-para-ong/ on August 17, 2017.

European Commission (n.d.) *Caribbean Area, Territorial co-operation.* [Web page]. Retrieved from http://ec.europa.eu/regional_policy/en/atlas/programmes/2014-2020/france/2014tc16rftn008

Ferro, V.G., Lemes P., Melo, A.S., and Loyola R. (2014) The reduced effectiveness of protected areas under climate change threatens Atlantic Forest Tiger Moths. *PLoS ONE 9(9): e107792. doi:10.1371/journal.pone.0107792*

Ferguson, E. (2011, December 6) Smuggled Jamaican parrots big hit at Vienna Zoo: Austrians bust international trade in endangered species. *The Jamaica Observer.* Retrieved from http://www.jamaicaobserver.com/news/Smuggled-Jamaican-Parrots-big- hit-at-Vienna-Zoo_10314363 on December 06, 2017.

Fitzpatrick, S. and Keegan, W. (2007) Human impacts and adaptations in the Caribbean Islands: An historical ecology approach. *Earth and Environmental Science Transactions of the Royal Society of Edinburgh*, 98(1), 29-45 doi:10.1017/S1755691007000096.

Flanders Marine Institute. (2018) *Maritime boundaries geodatabase: Maritime boundaries and exclusive economic zones (200NM), version 10.* retrieved from https://doi.org/10.14284/312

Food and Agriculture Organization (FAO). (2001) *Land resources information systems in the Caribbean: Proceedings of a sub-regional workshop.* World Soil Resources Reports 95.1. Rome: FAO.

Food and Agriculture Organization (FAO). (2006a) *Global forest resources assessment 2005: Progress towards sustainable forest management.* FAO Forestry Paper 147 (FRA 2005) Rome: FAO. 284

Food and Agriculture Organization (FAO). (2006b) *Global forest resources assessment 2005: Report on fires in the Caribbean and Mesoamerican regions.* Fire Management Working Paper 12 FAO. Rome, Italy. 40 pp. Rome: FAO. http://www.fao.org/docrep/009/j7568e/j7568e00.htm

Food and Agriculture Organization (FAO). (2006c) *Regional review on aquaculture development. Latin America and the* Caribbean - 2005. FAO Fisheries Circular No. 1017/1. FIRI/C1017/1 Rome: FAO.

Food and Agriculture Organization (FAO). (2014a) *Forests and climate change in the Caribbean.* Rome: FAO. Retrieved from http://www.fao.org/documents/card/en/c/c34802da-3b5c-4c32-998b- 1ee2ad750ade/ on September 3, 2017.

Food and Agriculture Organization (FAO). (2014b) *The sustainable intensification of Caribbean fisheries and aquaculture.* Rome: FAO. Retrieved from http://www.fao.org/3/a-i3932e.pdf

Food and Agriculture Organization (FAO) (2015) *The global forest resources assessment.* Desk Reference. Rome: FAO. Retrieved from http://www.fao.org/3/a-i4808e.pdf

Food and Agriculture Organization (FAO) (2016) *The Caribbean must prepare for increased drought due to climate change.* [Web page]. Rome: FAO. Retrieved from. http://www.fao.org/americas/noticias/ver/en/c/419202/

Food and Agriculture Organization/Global Environment Facility (FAO/GEF) (n.d.) *FAO/GEF project document climate change adaptation in the Eastern Caribbean fisheries sector (CC4FISH)* FAO project ID: 621550 GEF/LDCF/SCCF Project ID: 5667.

Franks, J., Johnson, D. and Ko, D. (2016) Pelagic Sargassum in the tropical North Atlantic. *Gulf and Caribbean Research* Vol 27, SC6-1. Retrieved from https://aquila.usm.edu/cgi/viewcontent.cgi?article=1511 &context=gcr on August 2, 2018

Freedom House. (2017a) *Freedom of the press 2017.* Washington, D.C.: Freedom House. Retrieved from https://freedomhouse.org/report/table- country-scores-fotp-2017.

Freedom House. (2017b) *Freedom of the press 2017 methodology.* [Web page]. Washington, D.C.: Freedom House. Retrieved from https://freedomhouse.org/report/freedom-press-2017-methodology

Freedom House. (2017c) *Freedom of the press regional trends.* Washington, D.C.: Freedom House. Retrieved from https://freedomhouse.org/report/freedom-press/freedom-press-2017#regional-trends

FSC (2017) *Forest stewardship council facts and figures.* July 4, 2017.

Bonn: Forest Stewardship Council. Retrieved from https://ic.fsc.org/file- download.facts-figu res-july-2017.a-2020.pdf.

Gage, I., and Edwards, E. (2003) *Creating a Jamaican spinal forest: Multistakeholder management and development.* Québec: XII World Forestry Congress. Retrieved from http://www.fao.org/docrep/ARTICLE/WFC/XII/0530-C3.HTM

Geoghegan, T. (2014) *Green economies in the Caribbean: Perspectives, priorities and an action learning agenda.* London: International Institute for Environment and Development.

Global Environment Facility (GEF) (2012a) *Project Identification Form (PIF) Conserving biodiversity in coastal areas threatened by rapid tourism and physical infrastructure development. Dominican Republic.* Washington, D.C.: GEF. Retrieved from https://www.thegef.org/sites/default/files/project_documents/PI MS%2520 4955%2520DR%2520PIF%2520final.pdf 285

Global Environment Facility (GEF) (2012b) *Project Identification Form (PIF) Conserving biodiversity and reducing habitat degradation in protected areas and their buffer zones. Saint Kitts and Nevis.* Washington, D.C.: GEF. Retrieved from https://www.thegef.org/sites/default/files/project_documents/RE V%2520 PIF_0.pdf

Global Environment Facility (GEF) (2017) *GEF-7 replenishment programming directions and policy agenda (prepared by the Secretariat)* Second Meeting for the Seventh Replenishment of the GEF Trust Fund. Washington, D.C.: GEF.

Global Water Partnership Caribbean (GWP-C) and Caribbean Community Climate Change Centre (CCCCC) (2016) *Caribbean water security and climate resilient development: A regional framework for investment.* Boodram N., Nichols, K.; Woolhouse G., and Walmsly N. (Eds.) Port-of- Spain: GWP-C.

Global Water Partnership Caribbean (GWP-C) (2014) *Achieving development resilient to climate change: A sourcebook for the Caribbean water sector.* Port-of-Spain: GWP-C.

Government of the Dominican Republic (2010) *Political Constitution of the Dominican Republic proclaimed on January 26th.*

Published in the Official Gazette No. 10561, January 26, 2010. Santo Domingo: Government of the Dominican Republic. Retrieved from http://observatorioserviciospublicos.gob.do/baselegal/constituci on2010.p df

Government of the Dominican Republic (2018) *Resolution 10-2018. Which provides the regulations for the management of buffer zones of the conservation units of the National System of Protected Areas (SINAP) of the Dominican Republic.* Santo Domingo: Government of the Dominican Republic.

Gómez Valenzuela, V., Bonilla, S., and Alpízar, F. (2014) *Economic valuation of the National System of Protected Areas of the Dominican Republic. Ministry of Environment and Natural Resources.* Ministry of Environment and Natural Resources, Global Environment Facility, United Nations Development Program.

Gore-Francis, J., (2013) *Antigua and Barbuda SIDS 2014 preparatory progress report.* Retrieved from https://sustainabledevelopment.un.org/content/documents/1049 240Antig ua and Barbuda final.pdf

Government of Antigua and Barbuda. (2014) *Antigua and Barbuda national strategic action plan 2014-2020.* St. Johns: Government of Antigua and Barbuda.

Government of Grenada (2000) *Biodiversity strategy and action plan.* St Georges: Ministry of Finance.

Government of Sint Maarten. (2017) *Minister Lee reports on Ministry activities Post Irma.* Philipsburg: Government of Sint Maarten. Retrieved from http://www.sintmaartengov.org/PressReleases/Pages/Minister-Lee- Reports-on-Ministry-Activities-Post-Irma.aspx.

Government of St. Vincent and the Grenadines (2010) *The fourth national biodiversity report of St. Vincent and the Grenadines to the UNCBD.* Kingstown: Ministry of Health and the Environment.

Gómez, G., and Díaz, R. (2001) Second Latin American Symposium on Forest Seeds: Situation of Forest Sector of the Dominican Republic. Santo Domingo.

Green Climate Fund (2016) *Projects and programmes. Sustainable Energy Facility for the Eastern Caribbean*. [Web page]. Retrieved from http://www.greenclimate.fund/-/sustainable-energy-facility-for-the- eastern-caribbean on November 18, 2017. 286 Griffith, M.D., and Ashe, J. (1993) Sustainable Development of coastal and marine areas in Small Islands Developing States: A basis for integrated coastal management. *Ocean and Coastal Management* 21(1993): 269-284.

Guingand, A. (2008) *Economic valuation of the Portland Bight Protected Area,* Jamaica. Lionel Town: Caribbean Coastal Area Management Foundation.

Gyory, J., Mariano, A., and Ryan, E. (2018) *The Caribbean Current. Ocean Surface Currents.* [Web page]. Retrieved from. http://oceancurrents.rsmas.miami.edu/caribbean/caribbean.html

Halpern, B.S., Walbridge, S., Selkoe, K.A., Kappel, C.V., Micheli, F., D'Agrosa, C., Bruno, J.F., Casey, K.S., Ebert, C., Fox, H.E., Fujita, R., Heinemann, D., Lenihan, H.S., Madin, E.M.P., Perry, M.T., Selig, E.R., Spalding, M., Steneck, R., and Watson, Steneck. R., and Watson, Lenihan, H.S., Madin, E.M.P., Perry, M.T., Selig, E.R., Spalding, M., Steneck, R., and Watson, R. (2008) A global map of human impact on marine ecosystems. *Science,* 319(5865), 948-952.

Hargreaves-Allen, V. (2010) *The economic valuation of the natural resources of Andros.* Washington, D.C.: Conservation Strategy Fund.

Hargreaves-Allen, V. (2011) *The economic value of ecosystem services in the Exumas Cays; threats and opportunities for conservation .* Washington, D.C.: Conservation Strategy Fund.

Harold, S., and Eckert, K. (2005) *Endangered Caribbean sea turtles: An educator's handbook.* Technical Report 3. Beaufort, North Carolina, USA: Wider Caribbean Sea Turtle Conservation Network (WIDECAST) Retrieved from www.widecast.org: http://www.widecast.org/Resources/Docs/Harold_and_Eckert_2 005_Cari b_Sea_Turtles_Educators_Handbook.pdf

Hawkins, J.P. and Roberts, C.M. (2004) Effects of artisanal fishing on Caribbean coral reefs. *Conservation Biology* 18(1): 215-226.

Hedges, S. (2008) At the lower size limit in snakes: two new species of threadsnakes (Squamata: Leptotyphlopidae: Leptotyphlops) from the Lesser Antilles. *Zootaxa*, 1841(1), 1-30.

Hedges, S. (2018, April 5) *Caribherp: Amphibians and reptiles of Caribbean Islands.* (www.caribherp.org) Philadelphia: Temple University.

Hedges, S., and Conn, C. (2012) A new skink fauna from Caribbean islands (Squamata, Mabuyidae, Mabuyinae) *Zootaxa*, 3288(2012), 4-244.

Heileman (Ed.) (2005) *Caribbean environment outlook. Special edition for the Mauritius International Meeting for the 10-year Review of the Barbados Programme of Action for the Sustainable Development of Small Island Developing States.* UNEP, CARICOM, University of the West Indies.

Heileman, S. and Corbin, C. (2006) Caribbean SIDS. In UNEP/GPA. The *state of the marine environment: Regional assessments* (pp 213-245) The Hague: UNEP/GPA.

Helmer, E. H., Ramos, O., del Mar López, T, M. Quiñones, M. and Diaz, W. (2002) Mapping the forest type and land cover of Puerto Rico, a component of the Caribbean Biodiversity Punto caliente. *Caribbean Journal of Science.* 38: 3-4.

Houck, O.A. *Environmental law in Cuba.* Retrieved from http://www.law.fsu.edu/docs/default-source/journals/jluel/previous-issues/volume-16-number-1.pdf?sfvrsn=4

Hudson, M.A. (2016) How we're saving the mountain chicken frog from one of the world's worst wildlife diseases. *The Conversation.* Retrieved from https://theconversation.com/how-were-saving-the-mountain-chicken-frog-from-one-of-the-worlds-worst-wildlife-diseases-64033 on October 5, 2017. 287

Hudson, M.A., Young R. P., Jackson, J., Wengel, P. O., Martin L., James A., Sulton M., Garcia G., Griffiths R. A., Thomas R., Magin C., Bruford M. W.and Cunningham, A. A. (2016) Dynamics and genetics of a disease- driven species decline to near extinction: Lessons for conservation. *Scientific Reports* 6 (30772)

Intergovernmental Science-Policy Platform on Biodiversity and

Ecosystem Services (IPBES) (2018) *Summary for policymakers of the regional assessment report on biodiversity and ecosystem services for the Americas of the Intergovernmental Science-Policy Platform on Biodiversity and Ecosystem Services.* Rice, J., Seixas, C. S., Zaccagnini, M.E, Bedoya-Gaitán, M., Valderrama, N., Anderson, C.B., Arroyo, M.T.K., Bustamante, M, Cavender-Bares, J., Diaz-de-Leon, A., Fennessy, S., Garcia Marquez, J. R., Garcia, K., Helmer, E.H., Herrera, B., Klatt, B., Ometo, J.P., Rodriguez Osuna, V., Scarano, F.R., Schill, S., and Farinaci, J.S. Bonn: IPBES Secretariat.

Intergovernmental Panel on Climate Change (IPCC) (2001) *Climate change 2001: The scientific basis.* Contribution of Working Group I to the Third Assessment Report of the Intergovernmental Panel on Climate Change. Cambridge: Cambridge University Press.

Intergovernmental Panel on Climate Change (IPCC) (2014) *Climate change 2013: The physical science basis.* Cambridge: Cambridge University Press. Retrieved from.
http://ebooks.cambridge.org/ref/id/CBO9781107415324.

International Labour Organisation (ILO) (2017) *Women in business and management: gaining momentum in Latin America and the Caribbean.* International Labour Office, Bureau for Employers' Activities (ACT/EMP) Geneva: International Labour Organisation. Retrieved from http://www.ilo.org/wcmsp5/groups/public/---ed_dialogue/- act_emp/documents/publication/wcms_579085.pdf.

International Monetary Fund (IMF) (2014) *Haiti: Poverty reduction strategy paper.* Washington, D.C.: IMF.

International Monetary Fund (IMF) (2017a) *Haiti: Request for disbursement under the rapid credit facility-press release; staff report; and statement by the executive director for Haiti.* IMF Country Report No. 17/315. Washington, D.C.: IMF.

International Monetary Fund (IMF) (2017b) *The Bahamas: Selected issues.* IMF Country Report No. 17/38. Washington, D.C.: IMF.

International Research Institute for Climate and Society (IRI) (2014) *Module 4: Introduction to hydroclimatology, part 1: Introduction to climate variability,* New York, NY, USA: IRI.

International Tropical Timber Organization (ITTO) (2008) *Developing forest certification: Towards increasing the comparability and acceptance of forest certification systems*. ITTO Technical Series No 29. Yokahama: ITTO.

Ionta, G.M., Judd, W.S., Skean, J.D., and McMullen, C.K. (2012) Two new species of Miconia sect. Sagraea (Melastomataceae) from the Macaya Biosphere Reserve, Haiti, and twelve relevant new species combinations. *Brittonia* 64: 61. Retrieved from. https://doi.org/10.1007/s12228-011 -9214-0

International Union for Conservation of Nature (IUCN) (2016) *A global standard for the identification of Key Biodiversity Areas, Version 1.0*. Gland, Switzerland: IUCN.

International Union for Conservation of Nature (IUCN) (2017a) *Table 1. Numbers of threatened species by major groups of organisms (19962017)* Gland, Switzerland: IUCN. Retrieved from http://cmsdocs.s3.amazonaws.com/summarystats/2017-1_Summary_Stats_Page_Documents/2017_1_RL_Stats_Table _1.pdf 288

International Union for Conservation of Nature (IUCN) (2017b, October 1) *IUCN Red List of Threatened Species. Version 2017.2*. Gland, Switzerland: IUCN. Retrieved from http://www.iucnredlist.org

International Union for Conservation of Nature (IUCN) (n.d.) *IUCN-CMP unified classification of direct threats. Guidance notes. Version 3.2*. Gland: IUCN Retrieved from http://s3.amazonaws.com/iucnredlist-newcms/staging/public/attachments/3127/dec_2012_guidance_t hreats_c lassification_scheme.pdf

Jackson, J. (1997) Reefs since Columbus. *Coral Reefs,* 16(1), S23-S32.

Jackson, J.B.C., Donovan, M.K., Cramer, K.L., and Lam, V.V. (Eds.) (2014) *Status and trends of Caribbean coral reefs: 1970-2012*. Gland, Switzerland: Global Coral Reef Monitoring Network.

Jaitman, L., and Torre I. (2017) The cost of crime in the Caribbean: The accounting method. In H. Sutton and I. Ruprah (Eds.), *Restoring paradise in the Caribbean: Combating violence with numbers* (pp. 181196) Washington, D.C.: Inter-American Development Bank.

Jamaica Conservation and Devlopment Trust (JCDT) (2018) *Blue and John Crow Mountains National Park.* [Web page]. Retrieved from https://www.jcdt.org.jm/bjcmnp/about/park-management/8-park

Jamaica Environment Trust (2013) Defining the boundary. *The Jetter* 1(6), 6.

James, R. (2014) *British Council Assessment of the civil society in Jamaica.* Prepared for the British Council by Research and Strategy Solutions Ltd. Kingston: British Council.

Jenkins, R.W.G., Jelden, D., Webb, G.J.W., and Manolis, S.C. (2004) *Review of crocodile ranching programs.* Conducted for CITES by the Crocodile Specialist Group of IUCN/SSC. January - April 2004. Sanderson, NT, Australia: IUCN-SSC Crocodile Specialist Group.

Jessop, D. (2016) *Recognising the value of Caribbean civil society.* Retrieved from http://www.caribbean-council.org/wp-content/uploads/2016/02/The-View-from-Europe-Feb21-Recognising- the-value-of-Caribbean-civil-society.pdf on June 2017.

John, L. (2005) *The potential of non-timber forest products to contribute to rural livelihoods in the Windward Islands of the Caribbean.* CANARI Technical Report No. 334. Laventille: Caribbean Natural Resources Institute.

Johnson, K., and Hedges, S. B. (1988) Three new species of Calisto from Southwestern Haiti (Lepidoptera: Nymphalidae: Satyrinae) *Tropical Lepidoptera, 9*(2), 45-53.

Joint Nature Conservation Committee (JNCC) (2007) *Invasive species in the UK Overseas Territories.* Peterborough, UK: JNCC.

Jonas, H. D, Barbuto, V., Jonas, H. C, Kothari, A., and Nelson, F. (2014) New steps of change: Looking beyond protected areas to consider other effective area based conservation measures. *Parks, 20*(2), 111-128.

Jones, F. (2015) *Ageing in the Caribbean and the human rights of older persons.* ECLAC - Studies and Perspectives Series - The

Caribbean. Santiago, Chile: United Nations Economic Commission for Latin American and the Caribbean.

Jones, I. C., Banner, J. L. and Mwansa, B. J. (1998) *Geochemical constraints on recharge and groundwater evolution: The Pleistocene aquifer of Barbados.* San Juan : Third International Symposium on Tropical Hydrology and Fifth Caribbean Islands Water Resources Congress. 289

Kairo, M., Ali, B., Cheesman, O., Haysom, K., and Murphy, S. (2003) *Invasive species threats to the Caribbean region: Report to The Nature Conservancy.* CAB International. Retrieved from http://www.bu.edu/scscb/working_groups/resources/Kairo-et-al-2003.pdf

Kaly, U., Pratt, C., and Howorth, R. (2002) *Towards managing environmental vulnerability in small island developing states (SIDS)* Suva: South Pacific Applied Geoscience Commission (SOPAC)

Kauffman, J. B., Heider, C., Norfolk, J., and Payton, F. 2014. Carbon stocks of intact mangroves and carbon emissions arising from their conversion in the Dominican Republic. *Ecological Applications,* 24(3), 518-527.

Keith, D.A., Rodriguez-Clark, J.P., Nicholson, E., Aapala, K., Alonso, A., Asmussen, M., Bachman, S., Basset, A., Barrow, E.G., Benson, J.S., Bishop, M.J., Bonifacio, R., Brooks, T.M., Burgman, M.A., Comer, P., Comin, F.A., Essl, F., Faber-Langendoen, D., Fairweather, P.G., Holdaway, R.,J., Jennings, M., Kingsford, R.T., Lester, R.E., Nally, R.M., McCarthy, M.A., Moat, J., Oliveira-Miranda, M.A., Pisanu, P., Poulin, B., Reagan, T.J., Riecken, U., Spalding, M.D., and Zambrano-Martínez, S. (2013) Scientific Foundations for an IUCN Red List of Ecosystems. *PLoS ONE,* 8(5) e62111. Retrieved from https://doi.org/10.1371/journal.pone.0062111

Kerchner, C. and Bonilla Duarte, S. (2014) The value of ecosystem services in the Quita Espuela and Guaconejo Scientific Reserves. In Bonilla-Duarte, S., *Promotion of a payment scheme for environmental services through the economic valuation of water resources in the Quita Espuela and Guaconejo Scientific*

Reserves, Dominican Republic. (pp. 13-22) Santo Domingo: Instituto Tecnológico de Santo Domingo.

Kiai, M. (2017*) Report of the Special Rapporteur on the rights to freedom of peaceful assembly and of association.* A/HRC/35/28. New York: United Nations Human Rights Council.

Kolher, G., and Hedges, S. (2016) A revision of the green anoles of Hispaniola with description of eight new species (Reptilia, Squamata, Dactyloidae) *Novitates Caribaea*, 9, 1-135.

Kristie, L, and Nealon, J. (2016) Dengue in a changing climate. *Environmental Research:* 151, 115-123.

Laloë, J.O., Esteban N., Berkel J. and Hays G.C. (2016) Sand temperatures for nesting sea turtles in the Caribbean: Implications for hatchling sex ratios in the face of climate change. *Journal of Experimental Marine Biology and Ecology,* 474, 92-99. Retrieved from.
http://dx.doi.org/10.1016/j.jembe.2015.09.015

Lang, J. (2003) Caveats for the AGRRA 'Initial Results' Volume. In J. Lang, *Status of coral reefs in the Western Atlantic. Results of initial surveys, Atlantic and Gulf Rapid Reef Assessment (AGRRA) Programme* (pp. xv- xx) Washington, DC, USA: Atoll Research Bulletin no.496. National Museum of Natural History, Smithsonian Institute.

Langhammer, P. F., Bakkar, M. I., Bennun, L. A., Brooks, T. M., Clay, R.P., Darwall, W., De Silva, N., Edgar, G.J., Eken, G., Fishpool, L.D.C., Fonseca, G.A.B. da, Foster, M.N., Knox, D.H., Matiku, P., Radford, E.A., Rodrigues, A.S.L., Salaman, P., Sechrest, W., and Tordoff, A.W. (2007) *Identification and gap analysis of key biodiversity areas: Targets for comprehensive protected area systems.* Gland: IUCN (Best Practice Protected Area Guidelines Series 15).

Lefebvre, V. (2017) *Evaluation ex-post du projet FEM (GEF) PID Établissement d'un système national d'aires protégées financièrement soutenable - Rapport final.* Retrieved from
https://info.undp.org/docs/pdc/Documents/HTI/PID58509_SNAP _Rappor
t%20evaluation%20finale_V%20Lefebvre_%20valide.pdf 290

Le Nouvelliste. (2017, July 1) 257 ONG interdits de fonctionnement en Haïti. *Le Nouvelliste*. Retrieved from. http://lenouvelliste.com/article/175871/257-ong-interdits-de-fonctionnement-en-haiti September 15, 2017.

Le Quesne, T., Matthews, J., Heyden, C., Wickel, A.J., Wilby, R., Hartmann, J., Pegram, G., Kistin, E., Blate, G., Freitas, G., Kimura de, L., Eliot, G., Carla; McSweeney, C. and Sindorf, (2010) *Flowing forward: Freshwater ecosystem adaptation to climate change in water resources management and biodiversity conservation.* Gland, Switzerland: WWF.

Lexadin. (n.d.) *Legislation* (Lexadin) [Web page]. Retrieved from https://www.lexadin.nl/wlg/legis/nofr/legis.php

Linardich, C. , Ralph, G., Carpenter, K., Cox, N., Robertson, D.R., Harwell, H., Acero P., A., Anderson Jr., W., Barthelat, F., Bouchereau, J.L., Brown, J.J., Buchanan, J., Buddo, D., Collette, B., Comeros-Raynal, M., Craig, M., Curtis, M., Defex, T., Dooley, J., Driggers III, W., Elfes Livsey, C., Fraser, T., Gilmore Jr., R., Grijalba Bendeck, L., Hines, A., Kishore, R., Lindeman, K., Maréchal, J.P., McEachran, J., McManus, R., Moore, J., Munroe, T., Oxenford, H., Pezold, F., Pina Amargos, F., Polanco Fernandez, A., Polidoro, B., Pollock, C., Robins, R., Russell, B., Sayer, C., Singh-Renton, S., Smith-Vaniz, W., Tornabene, L., Van Tassell, J., Vié, J.C., and Williams, J.T (2017) *The Conservation Status of Marine Bony Shorefishes of the Greater Caribbean.* Gland, Switzerland: IUCN.

Lips, K.R., Brem, F., Brenes, R., Reeve, J.D., Alford, R.A., Voyles, J., Carey, C., Livo, L., Pessier, A.P., and Collins, J.P. (2006) Emerging infectious disease and the loss of biodiversity in a Neotropical amphibian community. *PNAS* 103: 3165-3170.

Loh, T., McMurray S.E., Henkel T.P., Vicente J., and Pawlik, J.R. (2015) Indirect effects of overfishing on Caribbean reefs: sponges overgrow reef-building corals. *PeerJ* 3:e901 https://doi.org/10.7717/peerj.901

Lowe, S., M. Browne, S. Boudjelas, and De Poorter, M. 2001. *100 of the world's worst invasive alien species. A selection from the global invasive species database.* Auckland: Species Survival

Commission of the World Conservation Union (IUCN),

Lugo, A.E. (2000) Effects and outcomes of Caribbean hurricanes in a climate change scenario. *The Science of the Total Environment* 262, 243-251.

Lugo, A.E. (2008) Visible and invisible effects of hurricanes on forest ecosystems: an international review. *Austral Ecology* 33: 368-398.

McElroy J.L., Potter B., and Towle, E. (1990) Challenges for sustainable development in small Caribbean islands. In W. Beller, P. d'Ayala. P. Hein (Eds.), *Sustainable development and environmental management of small islands* (pp. 299-316) UNESCO/Man and the Biosphere Series. Paris: Parthenon Publishing Group.

McGuire, G. (2016) *Barriers to identification an implementation of energy efficiency mechanisms and enhancing renewable energy technologies in the Caribbean.* Port of Spain: ECLAC.

Malena, C. (2015) *Improving the measurement of civil space.* London: Open Society Foundation.

Marsh, S. (2017a, July 22) Cuba seeks to revive mining sector with new lead and zinc mine. *Reuters.* https://www.reuters.com/article/us-cuba-   mining/cuba-seeks-to-revive-mining-sector-with-new-lead-and-zinc-mine-idUSKBN1A70K3. Accessed 15 September 2017.

Marsh, S. (2017b, August 1) Communist-run Cuba puts brakes on private sector expansion. *Reuters.* https://www.reuters.com/article/us-cuba-privatesector/communist-run-cuba-puts-brakes-on-private-sector-expansion-idUSKBN1AH4SL. Accessed 15 Sept. 2017.
291

Martínez Rivera, C.C. and Rodríguez Plaza, C. (2015) *The amphibians of southern Hispaniola: their natural history and conservation. Diversity and conservation status of amphibians. Field guide.* Philadelphia Zoo, ComuEcoMedia, Critical Ecosystem Fund Alliance, Museo Natural de Historia Natural de Santo Domingo, Société Audubon Haïti and Grupo Jaragua. Retrieved from

http://www.conservaciondeanfibios.org/uploads/2/1/6/1/2161519
2/anfibio s_from_south_of_the_hispaniola.pdf

Massol González, A., González, E., Massol Deyá, A., Deyá, T., and
Geoghegan. T. (2006) *Bosque del Pueblo, Puerto Rico: How a
fight to stop a mine ended up changing forest policy from the
bottom up.* Policy That Works for Forests and People. No. 12.
London: International Institute for Environment and
Development.

Masters, G. and Norgrove, L. (2010) *Climate change and invasive
alien species.* CABI Working Paper 1, 30 pp. Wallingford: CABI.

Maunder, M., Leiva, A., Santiago-Valentín, E., Stevenson, D.W.,
Acevedo- Rodriguez, P., Meerow, A.W., Mejía, M., Clubbe, C.,
and Francisco- Ortega, J. (2008) Plant conservation in the
Caribbean Island Biodiversity Hot Spot. *The Botanical Review,*
74(1), 197-207.

Medeiros, D., Hove, H., Keller, M., Echeverría, D., and Parry, J. E.
(2011) *Review of current and planned adaptation action: the
Caribbean.* Report for the International Institute for Sustainable
Development.

Mercer. J., Kelman, I. Alfthan B. and Kurvits T. (2012) Ecosystem-
based adaptation to climate change in Caribbean small island
developing states: Integrating local and external knowledge.
Sustainability 4(8):1908-1932. Retrieved from
http://www.mdpi.com/2071- 1050/4/8/1908/htm on 15 December
2016.

Millennium Ecosystem Assessment (2005) *Ecosystems and human
wellbeing-Synthesis.* Washington, D.C.: Island Press. Retrieved
from
http://www.millenniumassessment.org/documents/document.35
6.aspx.pd f

Miloslavich, P., Díaz, J., Klein, E., Alvarado, J., Diaz, C., Gobin, J.,
Escobar- Briones, E., Cruz-Motta, J.J., Weil, e., Cortés, J.,
Bastidas, A. C., Robertson, R., Zapata, F., Martín, A., Castillo, J.,
Kazandjian, A., and Ortiz, M. (2010) Marine biodiversity in the
Caribbean: Regional estimates and distribution patterns. *PLoS
ONE,* 5(8) Retrieved from

https://doi.org/10.1371/journal.pone.0011916

Ministry of Economy, Planning and Development (2012) *Law 1-12 National Development Strategy 2030.* Santo Domingo: Ministry of Economy, Planning and Development of the Dominican Republic.

Ministry of Environment and Natural Resources. (2011a) *Regulation for the declaration of private protected areas or conservation.* Santo Domingo: Ministry of Environment and Natural Resources of the Dominican Republic.

Ministry of Environment and Natural Resources. (2011b) *Estrategia nacional de conservación y uso sostenible de la biodiversidad y plan de acción 2011-2020 (ENBPA)* Santo Domingo: Ministerio de Medio Ambiente y Recursos Naturales de la República Dominicana.

Ministry of Environment and Natural Resources (2012) *Los Haitises National Park Management Plan 2012-2017.* Santo Domingo: Ministry of Environment and Natural Resources of the Dominican Republic. 292

Ministry of Environment and Natural Resources (2014) *Plan de manejo Parque Nacional Montaña La Humeadora 2014-2019.* Santo Domingo: Ministry of Environment and Natural Resources of the Dominican Republic.

Ministry of Environment and Natural Resources. (2015) *Regulation of co-management of protected areas in the Dominican Republic .* Santo Domingo: Ministry of Environment and Natural Resources of the Dominican Republic.

Ministry of Finance and Economic Affairs (2013) *Medium term growth and development strategy 2013 - 2020.* Bridgetown: Government of Barbados.

Mittermeier, R.A., Robles-Gil, P., Hoffmann, M., Pilgrim, J.D., Brooks, T.B., Mittermeier, C.G., Lamoreux, J.L. and Fonseca, G.A.B. (2004) *Punto calientes Revisited: Earth's biologically richest and most endangered terrestrial ecoregions.* Mexico D.F.: CEMEX.

Morton, M. (2009) *Management of critical species on Saint Lucia: Species profiles and management recommendations.* Technical Report No. 13 to the National Forest Demarcation and Bio-

Physical Resource Inventory, Helsinki: FCG International Ltd.

Mumby, P.J. and Fitzsimmons, J.F.M. (2014) *Towards reef resilience and sustainable livelihoods: A handbook for Caribbean coral reef managers.* Exeter: University of Exeter.

Munro-Knight, S. (2013) *Caribbean civil society and sustainable Caribbean development: Initial discussion paper for the 3rd Conference on Small Island Developing (SIDS 2014)* Caribbean Consultative Working Group. Bridgetown: Caribbean Policy Development Centre.

Myers, N., Mittermeier, R., Mittermeier, C., da Fonseca, G., and Kent, J. (2000) Biodiversity hotspot for conservation priorities. *Nature, 403(6772), 853.* doi:10.1038/35002501

Myers, R., O'Brien, J., Mehlman D., and Bergh, C. (2004a) *Fire management assessment of the highland ecosystems of the Dominican Republic.* GFI publication no. 2004-2. Arlington, VA: The Nature Conservancy.

Myers, R., Wade, D. and Bergh, C. (2004b) *Fire management assessment of the Caribbean Pine (Pinus caribea) forest ecosystems on Andros and Abaco Islands, Bahamas.* GFI publication no. 2004-1. Arlington, VA: The Nature Conservancy.

Narcisse, C. (2017, September 5) CSOs and governance in Jamaica and the Caribbean. *Telephone interview* with Nicole A. Brown.

National Environment and Planning Agency. (2016) *National strategy and action plan on biological diversity in Jamaica 2016-2021.* Kingston: National Environment and Planning Agency.

National Oceanic and Atmospheric Administration (NOAA) (2015, October 8) *NOAA declares third ever global coral bleaching event. Bleaching intensifies in Hawaii; high ocean temperatures threaten Caribbean corals.* [Web page]. Retrieved from http://www.noaanews.noaa.gov/stories2015/100815-noaa-declares-third-   ever-global-coral-bleaching-event.html   on   8 October 2015.

National Oceanic and Atmospheric Administration (NOAA) (n.d.) *Global coral bleaching 2014-2017: Status and an appeal for observations.* [Web page]. Retrieved from https://coralreefwatch.noaa.gov/satellite/analyses_guidance/glo

bal_coral _bleaching_2014-17_status.php

National Renewable Energy Laboratory (2015) *Energy snapshot Dominican Republic.* Energy Transitions Initiative Islands. Washington, D.C.: U.S. Department of Energy, Office of Energy Efficiency and Renewable Energy. 293

NDP Secretariat. (2017) *Vision 2040: National development plan of The Bahamas (Second working draft)* Nassau: Government of the Commonwealth of The Bahamas.

Nieto-Blázquez, M., Antonelli, A., and Roncal, J. (2017) Historical biogeography of endemic seed plant genera in the Caribbean: Did GAARlandia play a role? *Ecology and Evolution,* 7(23), 10158-10174. doi:https://doi.org/10.1002/ece3.3521

Now Grenada. (2017, May 28) *Grand Anse Marine Protected Area established* [Web page]. Retrieved from. http://www.nowgrenada.com/2017/05/grand-anse-marine-protected- area-gampa-established/

Now Grenada (2018, January 22) *Grenada launches its newest protected area: Grand Anse.* [Web page]. Retrieved from. http://www.nowgrenada.com/2018/01/grenada-launches-its-newest- marine-protected-area-area-grand-anse/

Oceans 5. (n.d.) *Grants database.* [Data file]. Retrieved from http://oceans5.org/grants/ on February 24, 2018.

Ogawa, S., Park, J., Singh, D., and Thacker, N. (2013) *Financial interconnectedness and financial sector reform in the Caribbean.* Washington, D.C.: International Monetary Fund. Retrieved from https://www.imf.org/external/pubs/ft/wp/2013/wp13175.pdf   on November 18, 2017.

Onestini, M. (2017) *Strengthening the operational and financial sustainability of the national protected area system - Terminal evaluation.*   Retrieved   from https://erc.undp.org/evaluation/evaluations/detail/8896

Organisation of Eastern Caribbean States (OECS) (2011) *Common tourism policy.* Castries: OECS Secretariat.

Palasi, J.P., Martinez, C., and Laudon, A. I. (2006J *Financements publics et biodiversité en outre-mer : Quelle ambition pour le développement durable ?* IUCN French National Committee

Pan American Health Organization/World Health Organization (PAHO/WHO) (2017, May 4) *Epidemiological Update: Cholera.* http://www.paho.org/hq/index.php?option=com_docmanandtask =doc_vie wandgid=39836andItemid=270andlang=en

Pantin, D., and Reid, V. (2005) *Economic valuation study: Action-learning project on incentives for improved watershed services in the Buff Bay/ Pencar Watershed.* CANARI Who Pays for Water Project Document No.2. Laventille: CANARI

Parra-Torrado, M. (2014) *Youth unemployment in the Caribbean.* Caribbean Knowledge Series. Washington, D.C.: World Bank. Retrieved from http://documents.worldbank.org/curated/en/7564314680126435 44/Youth -unemployment-in-the-Caribbean.

Petit, J., and Prudent, G. (2010) *Climate change and biodiversity in the European Union overseas entities.* Gland, Switzerland and Brussels, Belgium: IUCN.

Pérez-Silva, J. B. y Hernández-Muñoz, A. (2022) New locality of *Tropidophis Spiritus* Hedges and Garrido, 1999 (Serpentes: Tropidophiidae) in Sancti Spíritus, Cuba. *Infociencia Magazine,* ISSN 1029-5186, Vol. 26, No.3. September 2022 - December, p. 1-7.

Planning Institute of Jamaica (2009) *Vision 2030 Jamaica: national development plan.* Kingston: Planning Institute of Jamaica.

Polunin, N., and Williams, I. (1999) *Reef condition and recovery in the Caribbean.* Marine Systems Research Group Technical Report 6. A Subject Review for the Department for International Development (DFID) NRSP LWI Project R6783. Government of the United Kingdom.

Population Reference Bureau (2008) *World population data sheet 2008.* Washington D.C.: Population Reference Bureau. http://www.prb.org/pdf08/08WPDS_Eng.pdf 294

Population Reference Bureau (2017) *World population data sheet 2017.* Washington D.C.: Population Reference Bureau Retrieved from https://assets.prb.org/pdf17/2017_World_Population.pdf

Posner, S., Michel, G. and J. R. Toussaint. (2010) *Haiti biodiversity and tropical forest assessment.* Washington, D.C.: USAID,

USDA Forest Service.

Pressey, R. L. (1994) Ad hoc reserves: forward or backward steps in developing representative reserve systems. *Conservation Biology*, 8, 662-668.

Proctor, C., Inman, S., Zeiger, J., and Graves, G. (2017) Last search for the Jamaican golden swallow (*Tachycineta e. euchrysea*) *The Journal of Caribbean Ornithology*, 30(1), 69-74. Retrieved from. www.birdscaribbean.org/jco

Puerto Rico Climate Change Council (PRCCC) (2013) *Puerto Rico's State of the Climate 2010-2013: Assessing Puerto Rico's social-ecological vulnerabilities in a changing climate.* Puerto Rico Coastal Zone Management Program. San Juan: Department of Natural and Environmental Resources, NOAA Office of Ocean and Coastal Resource Management.

Quarless, D. (2015) *Ageing in the Caribbean and the rights of older persons.* Briefing Note by ECLAC Caribbean Chief. Retrieved from https://www.cepal.org/en/notas/ageing-caribbean-and-rights-older- persons

Quarless, D. (2017) *Caribbean economies.* Briefing note by ECLAC Caribbean Chief. Retrieved from. https://www.cepal.org/en/notes/caribbean-economies

Quiñones Rosado, R. E. (2002) *Regulation of civil society organizations (CSOs): Legislation and proposals.* Santo Domingo: Instituto Tecnológico de Santo Domingo, Inter-American Development Bank.

Quiroga, R.; Perazza, Maria, C.; Corderi, D.; Banerjee, O.; Cotta, J.; Watkins, G. G.; López Sancho, J.L. (2016) *Environment and biodiversity: Priorities for protecting natural capital and competitiveness in Latin America and the Caribbean.* Retrieved from https://publications.iadb.org/handle/11319/7885#sthash.WrChH 0vR.dpuf

Rawwida Baksh and Associates. (2016) *Country gender assessments (CGAs) synthesis report.* Bridgetown: Caribbean Development Bank. Retrieved from http://www.caribank.org/wp-content/uploads/2016/05/SynthesisReportCountryGenderAsses

sment.pd f

Renard, Y., and Borobia, M. (2015) *Terminal evaluation of the UNEP Project Caribbean Biological Corridor.* Evaluation Report, UNEP.

Reynolds, R., Puente-Rolón, A., Geneva, A., Aviles-Rodriguez, K., and Herrmann, N. (2016) Discovery of a remarkable new boa from the Conception Island Bank, Bahamas. *BREVIORA,* 519(1), 1-19. Retrieved from http://dx.doi.org/10.3099/brvo-549-00-1-19.1

Roberts, C.M., McClean, C.J., Veron, J.E.N., Hawkins, J.P., Allen, G.R., McAllister, D.E., Mittermeier, C.G., Schueler, F.W., Spalding, M., Wells, F., Vynne, C., and Werner, T.B. (2002) Marine biodiversity hotspot and conservation priorities for tropical reefs. *Science*, 295, 1280-1284.

Rojas. E., Wirshafter, R.M., Radke, J., and Hosier, R. (1988) Land conservation in small developing countries: computer assisted studies in Saint Lucia. *Ambio* 17: 282-288.

Scalley, T.H. (2012) Freshwater resources in the insular Caribbean: an environmental perspective. *Caribbean Studies* 40(2), 63-93.

Schwartz, M. (1999) Choosing the appropriate scale of reserves for conservation. *Annual Review of Ecology and Systematics,* 30(1), 83108.

Seavy, N.E., Gardali, T., Golet, G.H., Griggs, F.T., Howell, C.A., Kelsey, R., Small, S.L., Viers, J.H. and Weigand, J.F. (2009) Why climate change makes riparian restoration more 295 mportant than ever: Recommendations for practice and research. *Ecological Restoration*, 27(3), 330-338.

Secretariat for the Convention on Biodiversity. (2018) *Country profiles.* [Web page]. Retrieved from https://www.cbd.int/countries/

Seddon, N., Hou-Jones, X., Pye T., Reid H., Roe D., Mountain, D. and Raza Rizv, A. (2016) *Ecosystem-based adaptation: A win-win formula for sustainability in a warming world?* IIED Briefing. London: International Institute for Environment and Development.

Simpson M.C.; Scott D.; Harrison M.; Silver, N., O'Keeffe, E., Sim, R., Harrison, S., Taylor, M., Lizcano, G., Rutty, M., Stager, H.,

Oldham, J., Wilson, M., New, M., Clarke, J., Day, O.J., Fields, N., Georges, J., Waithe, R. and McSharry, P. (2010) *Quantification and magnitude of losses and damages resulting from the impacts of climate change: Modelling the transformational impacts and costs of sea level rise in the Caribbean (Summary Document)* Barbados: United Nations Development Programme (UNDP)

Singh, R. (2017, June 26) *New benchmarks to assess press freedom in the Caribbean. Guyana Chronicle.* Retrieved from https://guyanachronicle.com/2017/06/26/new-benchmarks-to-assess- press-freedom-in-caribbean on September 15, 2017.

Skean, J.D. (2000) Mecranium juddii (Melastomataceae: Miconieae), a new species from the Massif de la Hotte, Haiti *Brittonia* 52: 191. Retrieved from https://doi.org/10.2307/2666511

Skerratt L.F., Berger L., Speare R., Cashins S., McDonald, K.R., Phillot, A.D., Hines, H.B., and Kenyon, N. (2007) Spread of chytridiomycosis has caused the rapid global decline and extinction of frogs. *EcoHealth* 4: 125-134.

Smith, M. L. (2004) Caribbean Islands. In Mittermeier, R.A., Gil, P.G., Hoffman, M., Pilgrim, J., Brooks, T.M., Mittermeier, C.G., Lamoreux, J., Da Fonseca, G.A.B., (Eds.), *Punto calientes revisited: Earth's biologically richest and most endangered terrestrial ecoregions.* Mexico DF, Mexico: CEMEX.

Smucker, G.R., Bannister, M., D'Agnes, H., Gossin, Y., Portnoff, M., Timyan, J., Tobias, S., and Toussaint, R. (2007) *Environmental vulnerability in Haiti: Findings and recommendations.* Washington, D.C.: USAID.

Smulders, F.O.H, Vonk, J.A., Engel, M. S. and Christianen M. J. J. A. (2017) Expansion and fragment settlement of the non-native seagrass Halophila stipulacea in a Caribbean bay, *Marine Biology Research,* 13:9, 967-974, DOI: 10.1080/17451000.2017.1333620

Soulé, M., and Terborgh, J. (1999) *Continental conservation: Scientific foundations of regional reserve networks.* Washington, D.C., USA: Island Press.

Spalding, H. (n.d.) *Cuba's evolving civil society.* [Web page]. NACLA. Retrieved from https://nacla.org/article/cuba%E2%80%99s-evolving- civil-society

Spalding, M., Green, E., and Ravilious, C. (2001) *World atlas of coral reefs.* California: UNEP World Conservation Monitoring Centre. University of California Press.

Springer, C. 2005. *Cost pricing for water production and water protection services in Jamaica: a situational analysis.* Impact Consultancy Services Incorporated on behalf of The Caribbean Natural Resources Institute (CANARI) and International Institute for Environment and Development.

Springer, S. (1950) An outline for a Trinidad shark fishery. *Proceedings of the Gulf and Caribbean Fisheries Institute second annual session* (pp. 1726) Coral Gables: University of Miami. 296

Stattersfield, A., Crosby, M., Long, A., and Wege, D. (1998) *Endemic Bird Areas of the world: priorities for biodiversity conservation.* Cambridge, U.K.: BirdLife International.

Stein, B. A., Staudt, A., Cross, M. S. Dubois, N. S., Enquist C., Griffis, R. Hansen, L. J. Hellmann, J. J., Lawler, J. J., Nelson, E. J., and Pairis. A. (2013) Preparing for and managing change: climate adaptation for biodiversity and ecosystems. *Frontiers in Ecology,* 11(9), 502-510.

Stephen, C. L., Reynoso, V. H., Collett, W. S., Hasbun, C. R., and Breinholt, J. W. (2013) Geographical structure and cryptic lineages within common green iguanas, *Iguana iguana. Journal of Biogeography,* 40(1), 50-62.

Stephenson, T.S., Vincent, L., Allen, T. Van Meerbeeck, C.J., McLean, N., Peterson, T.C., Taylor, M.A.; Aaron-Morrison, A. P., Auguste, T., Bernard, D., Boekhoudt, J.R.I., Blenman, R.C., Braithwaite, G.C., Brown, G., Butler, M., Cumberbatch, C.J.M., Etienne-Leblanc, S., Lake, D.E., Martin, D.E., McDonald, J.L., Zaruela, M.O., Porter, A.O., Ramirez, M.S., Tamar, G.A., Roberts, B.A., Mitro, S.S., Shaw, A., Spence, J.M., Winter, A., and Trotman A.R. (2014) Changes in extreme temperature and precipitation in the Caribbean region, 1961-2010. *International Journal of Climatology,* 34(9), 2957-2971.

Suárez, A.G., Garraway, E., Vilamajó, D., Mujica, L., Gerhartz, J., Capote, R. T. and Blake, N. (2008) *Climate change impacts on terrestrial biodiversity in the insular Caribbean.* Report of Working Group III, Climate. Port of Spain: Caribbean Natural Resources Institute (CANARI).

Sutton, H. and Alvarez, L. (2017) Who is most likely to be a victim of Crime? In H. Sutton and I. Ruprah (Eds.), *Restoring paradise in the Caribbean: Combating violence with numbers,* (pp.33- 2009) Washington, D.C.: Inter-American Development Bank.

Heather Sutton, H., Jaitman, L. and Khadan, J. (2017) Violence in the Caribbean: Cost and impact. In T. Alleyne, Î. Otker, U. Ramakrishnan, and K. Srinivasan (Eds.), *Unleashing growth and strengthening resilience in the Caribbean.* (pp. 329 - 345) Washington, DC: International Monetary Fund.

Sutton, H., Ruprah, I., and Pecha. C. (2017) The Effects of Crime on Economic Growth, Tourism, Fear, Emigration, and Life Satisfaction. In H. Sutton and I. Ruprah (Eds.), *Restoring paradise in the Caribbean: Combating violence with numbers,* (pp.197- 209) Washington, D.C.: Inter-American Development Bank.

Taylor. M. (2017) *Climate change and the Caribbean - The take away messages.* Video File. Kingston: Panos Caribbean. Retrieved from https://www. youtube.com/watch?v=G15y6UmBfFI

The Daily Observer. (2017, May 22) Yida agreement deemed a sign of desperation. *The Daily Observer.* Retrieved 31 January 2018 from: https://antiguaobserver.com/yida-agreement-deemed-a-sign-of- desperation/.

The Economist Intelligence Unit Limited (2015) *Private sector development in the Caribbean: A regional overview.* London: The Economist Intelligence Unit Limited.

The Economics of Ecosystems and Biodiversity (TEEB) (2018) *Ecosystem services.* [Web page]. Retrieved from http://www.teebweb.org/resources/ecosystem-services/ on April 4, 2018.

The MacArthur Foundation (n.d.) *Our work.* [Web page]. Retrieved from https://www.macfound.org/programs/conservation/ February 24, 2018.

The Mohamed bin Zayed Species Conservation Fund (n.d.)

Supported projects. [Web page] Retrieved from https://www.speciesconservation.org on February 24, 2018. 297

The Nature Conservancy (2015) *CEPF Final project completion report: A campaign to promote private sector sustainable finance mechanisms to support Jamaica's protected areas system. Project Report.*

The Nature Conservancy (2016) *2016 Caribbean impact report.* Coral Gables: The Nature Conservancy.

The Nature Conservancy (n.d.) *Caribbean. The Caribbean Challenge Initiative.* Retrieved from

https://www.nature.org/ourinitiatives/regions/caribbean/caribbean- challenge.xml

The World Bank (2012) *Expanding financing for biodiversity conservation experiences from Latin America and the Caribbean.* Latin America and Caribbean Region. Environment and Water Resources Occasional Paper Series. Washington, D.C.: The World Bank.

http://www.worldbank.org/content/dam/Worldbank/document/LAC- Biodiversity-Finance.pdf

The World Bank. (2014) *Saint Lucia disaster vulnerability reduction project* (English) Washington, D.C.: The World Bank.

The World Bank. (2016) *Jamaica disaster vulnerability reduction project* (English) Washington, D.C.: The World Bank. http://documents.worldbank.org/curated/en/282911467999736588/pdf/P AD1233-R2016-0008-1-Box394844B-OUO-9.pdf

The World Bank (2017a) *List of economies June 2017.* [Data file]. Retrieved from http://databank.worldbank.org.

The World Bank. (2017b) *World Bank open data.* [Data file]. Retrieved from https://datacatalog.worldbank.org between August and November, 2017.

The World Bank. (2017c) *World development indicators.* Washington, D.C.: The World Bank.

The World Bank. (2017d) *Global economic prospects, June 2017: A fragile recovery.* Washington, D.C.: World Bank. doi: 10.1596/978-1-4648-10244.

http://documents.worldbank.org/curated/en/503571468301159

47/Saint- Lucia-Disaster-Vulnerability-Reduction-Project

The World Bank. (2017e) *Organization of Eastern Caribbean States - Caribbean regional oceanscape project* (English) Washington, D.C.: World Bank Group. http://documents.worldbank.org/curated/en/6240815052299280 94/Organ ization-of-Eastern-Caribbean-States-Caribbean- Regional-Oceanscape- Proje

The World Bank. (2017f) St. Vincent and the Grenadines - Regional disaster vulnerability reduction project: restructuring (English) Washington, D.C.: World Bank Group. http://documents.worldbank.org/curated/en/1247314875596239 84/St- V ncent-and-the-Grenadines-Regional-Disaster- Vulnerability-Reduction- Project-restructuring

Theile, S., Steiner, A. and Kecse-Nagy, K. (2004) *Expanding borders: New challenges for wildlife trade controls in the European Union.* Brussels: TRAFFIC Europe.

Thompson (1944) *The fisheries of British Honduras.* Development and Welfare in the West Indies, Bulletin 21.

Timyan, J. (2011) *Key Biodiversity Areas of Haiti.* Port au Prince: Société Audubon Haïti. Retrieved from http://audubonhaiti.org/wordpress/wp- content/uploads/2012/09/KEY-BIODIVERSITY-AREAS-OF- HAITI_FINAL.pdf.

Tornabene, L., Robertson, D., and Baldwin, C. (2016) *Varicus lacerta, a new species of goby (Teleostei, Gobiidae, Gobiosomatini, Nes* subgroup) from a mesophotic reef in the southern Caribbean. *ZooKeys,* 596, 143156. doi:10.3897/zookeys.596.8217 298

Toussaint, R. (n.d.) *Profile of the Haiti national biodiversity strategy and action plan with implications for bi-national actions with the Dominican Republic.* Port au Prince: NEAP SecretariatfMinistry of Environment.

Towle, J. A., Moody, W. S., and Randall, A. (2010) *Philanthropy, civil society, and law in the Caribbean: A preliminary overview of the legal framework supporting philanthropy and the non-profit sector in the insular Caribbean.* A report of the Caribbean Philanthropy Network. St. Thomas: USVI.

Towle, J. (2017) *Implementation of the Virgin Islands Non-Profit (NPO) Act, 2012: Implications for NPOs, civil society and philanthropy in the Virgin Islands.* Caribbean Philanthropy Network and Community Foundation of the Virgin Islands.

United Nations. (2015) *The Millennium Development Goals report 2015- Regional backgrounder: Latin America and the Caribbean.* Retrieved from www.un.org/millenniumgoals/2015_MDG_Report/pdf/backgrounders/MD G%202015%20PR%20Bg%20LAC.pdf November 18, 2017.

United Nations. (2016) *World statistics pocketbook 2016 edition, V (2)* New York, NY: Department of Economic and Social Affairs Statistics Division, United Nations.

United Nations Convention to Combat Desertification. (2017) Global land outlook. First edition. Bonn: UNCCD.

United Nations Department of Economic and Social Affairs, Population Division. (2014) *World urbanization prospects: The 2014 revision, highlights.* (ST/ESA/SER.A/352) New York: United Nations.

United Nations Development Programme (UNDP) (2016a) *Human development report: Human development for everyone.* New York: United Nations Development Programme.

United Nations Development Programme (UNDP) (2016b) *Regional human development report 2016: Caribbean. Multidimensional progress: Human resilience beyond income.* Retrieved from http://hdr.undp.org/sites/default/files/undp_bb_chdr_2016.pdf on November 18, 2017.

United Nations Educational, Scientific and Cultural Organisation (UNESCO) (2006) *The use of desalination plants in the Caribbean.* IHP-LAC Technical Papers, No. 5. Caribbean Environmental Health Institute.

United Nations Educational, Scientific and Cultural Organisation (UNESCO (2018) *World Heritage List Blue and John Crow Mountains* [Web page]. Retrieved from http://whc.unesco.org/en/list/1356

United Nations Educational, Scientific and Cultural Organisation

(UNESCO. (n.d.) *Free media contribute to good governance, empowerment and eradicating poverty*. [Web page]. Retrieved from
http://www.unesco.org/new/en/unesco/events/prizes-and-celebrations/celebrations/international-days/world-press-freedom-day/2014-themes/free-media-contribute-to-good-governance-empowerment-and-eradicating-poverty/ on September 18, 2017.

United Nations Environment Programme (UNEP) (2002) *Global Environment Outlook* 2002. London: Earthscan.

United Nations Environment Programme (UNEP) (2004a) Villasol, A. and Beltrán, J. *Caribbean islands, GIWA regional assessment 4.* Fortnam, M. and P. Blime (Eds.) Kalmar, Sweden: University of Kalmar.

United Nations Environment Programme (UNEP) (2004b) Bernal, M.C., Londoño, L.M., Troncoso, W., Sierra-Correa, P.C. and Arias-Isaza, F.A. (Eds.), *Caribbean Sea/small islands, GWA Regional Assessment 3a.* Kalmar, Sweden: University of Kalmar.

United Nations Environment Programme (UNEP) (2008) *Climate change in the Caribbean and the challenge of adaptation.* United Nations Environment Programme, Regional Office for Latin America and the Caribbean.

United Nations Environment Programme (UNEP) (2011) *Towards a green economy: Pathways to sustainable development and poverty eradication. A synthesis for policy makers.* Nairobi: UNEP. Retrieved from https://sustainabledevelopment.un.org/content/documents/126G ER_synt hesis_en.pdf

United Nations Environment Programme (UNEP) (2014) *Integrating water, land and ecosystems management in Caribbean Small Island Developing States project document. Project no. GFL/4932.* Nairobi: UNEP.

United Nations Environment Programme (UNEP) (2016a) *Enhancing cooperation among the seven biodiversity related agreements and conventions at the national level using national biodiversity strategies and action plans.* Nairobi: UNEP.

United Nations Environment Programme (UNEP) (2016b) *Environment in the 2030 agenda for LAC. Elements for session 1: Sustainable development.* Twentieth Meeting of the Forum of Ministers of Environment of Latin America and the Caribbean, Cartagena de Indias, Colombia. *UNEP/LAC-XX/3 Rev.1.* New York: UNEP.

United Nations Environment Programme (UNEP) (2016c) *Haiti South Department forest energy supply chains.* UNEP Haiti, September 2016. Nairobi: UNEP.

United Nations Environment Programme - Caribbean Regional Coordinating Unit (UNEP RCU) (2001) *Coastal Zone Management, Environmental Issues in the Caribbean, Caribbean Environment Programme.* [Web page]. Retrieved from http://www.cep.unep.org/issues/czm.html

United Nations Environment Programme/Convention on Biological Diversity (UNEP/CBD) (2009) *Capacity-development workshop for the Caribbean region on national biodiversity strategies and action plans and mainstreaming of biodiversity and integration of climate change.* Port-of-Spain, Trinidad and Tobago: UNEP/CBD.

United Nations Environment Programme/Regional Office on Latin America and the Caribbean (UNEP/ROLAC) (2012) *Bi-annual report on the implementation of the UNEP/EC project entitled "The demarcation and establishment of the Caribbean biological corridor (CBC): as a framework for biodiversity conservation, environmental rehabilitation and development of livelihoods."* Panama City: UNEP/ROLAC. Retrieved from https://cbcinfo.files.wordpress.com/2012/09/cbc-final-report_050812.pdf

United Nations Environment Programme -World Conservation Monitoring Centre UNEP-(WCMC) (2016) *The state of biodiversity in Latin America and the Caribbean: A mid-term review of progress towards the Aichi Targets.* Cambridge, UK: UNEP-WCMC.

United Nations Programme on HIV and AIDS (UNAIDS) (2017) *Ending AIDS: Progress.* Geneva: UNAIDS.

United Nations Statistics Division. (2018) *2. Population, latest available census and estimates (2015 - 2016)* Retrieved from https://unstats.un.org/unsd/demographic-social/products/vitstats/seratab2.pdf.

United States Department of Agriculture (USDA) Caribbean Climate Hub (n.d.) *Drought effects on forests and rangelands in the US Caribbean.* [Web page].retrieved from 300

https://www.climatehubs.oce.usda.gov/hubs/caribbean/topic/drought -effects- forests-and-rangelands-us-caribbean on October 30, 2017.

Varty, N. (2016) *Terminal evaluation of the UNEP/GEF Project: "Project for Ecosystem Services (ProEcoServ)"* [GEF project ID: 3807]. Nairobi, Kenya: UNEP Evaluation Office.

Vaslet, A., and Renoux, R. (2016) *Regional ecosystem profile - Caribbean region. EU outermost regions and overseas countries and territories* . BEST Service Contract 07.-3-7.2013/666363/SER/B2. Brussels: European Commission.

Vergés, A., Steinberg, P.D., Hay, M.E., Poore, A.G.B., Campbell, A. H., Ballesteros, E. Heck, Jr. K.L., Booth, D.J. and Coleman, M.A. (2014) The tropicalization of temperate marine ecosystems: Climate-mediated changes in herbivory and community phase shifts. *Proceedings of the Royal Society B.* 281. (1789) Proc. R. Soc. B 281: 20140846.

Vidal, J. (2014, February 10) Wind of change sweeps through energy policy in the Caribbean. *Guardian Newspaper* (online version) Retrieved from https://www.theguardian.com/global-development/poverty-matters/2014/feb/10/wind-of-change-energy-policy-caribbean

Watson, C., Patel, S., Durand, A., and Schalatek. L. (2016) *Climate finance briefing: Small Island Developing States.* Climate Finance Fundamentals 12. London: Overseas Development Institute and Washington, D.C.: Heinrich Boll Stiftung North America.

Watson, J., Cross, M., Rowland, E., Joseph, L.N., Rao, M., and Seimon, A. (2011) Planning for species conservation in a time of climate change. In J. Blanco and H. Kheradmand (Eds.), *Climate change - Research and technology for adaptation and mitigation.* Rijeka: InTech

Weaver, D. B. (1993) Ecotourism in the small island Caribbean. *GeoJournal, 31(4),* 457-465.

Webson, W. A. (2010) *Philanthropy, civil society, and NGOs in the Caribbean: An overview of the dimensions of the NGO/civil society sector in the insular, English-speaking Caribbean. A report of the Caribbean Philanthropy Network.* St. Thomas: USVI: Caribbean Philanthropy Network.

Western Central Atlantic Fishery Commission (2017) *Review of the state of fisheries in FAO Area 31.* Eighth Session of the Scientific Advisory Group (SAG). Merida, Mexico, 3-4 November 2017. Retrieved from http://www.fao.org/fi/static-media/MeetingDocuments/WECAFC/WECAFC17/3Reve.pdf.

WIDECAST. (2018) *Conservation status - IUCN Red List.* Retrieved from http://www.widecast.org/conservation/iucn-red-list/

Wielgus, J., Cooper, E, Torres R. and Burke, L. (2010) *Coastal capital: Dominican Republic. Case studies on the economic value of coastal ecosystems in the Dominican Republic. Working Paper.* Washington, DC: World Resources Institute.

Wikipedia (2016) *Papa Bois.* [Web page]. Retrieved from. https://en.wikipedia.org/w/index.php?title=Papa_Boisandoldid=7 3146067 7 on February 23, 2018.

Wikipedia (n.d.) *Carbon dioxide equivalent.* [Web page]. Retrieved from https://en.wikipedia.org/w/index.php?title=Carbon_dioxide_equi valentand oldid=828621821 on March 3, 2018

Wiley, J. W., and Wunderle, J. M. (1993) The effects of hurricanes on birds, with special reference to Caribbean islands. *Bird Conservation* 3: 319349. 301

Wilkinson, C., and Souter, D. (2008) *Status of Caribbean Coral reefs after bleaching and hurricanes in 2005.* Townsville: Global Coral Reef Monitoring Network, and Reef and Rainforest Research Centre.

Wilson, E. O. (1992) *The diversity of life.* Harvard University Press.

Wilson, S., Sagewan-Alli, I. and Calatayud A. (2014) *The ecotourism industry in the Caribbean. A value chain analysis.* IDB-TN-710. Inter-American Development Bank.

Windsor Research Centre. (2014) *Cockpit Country watersheds Jamaica.* Retrieved from http://www.cockpitcountry.com/watersheds.html

Woods, C.A.and Sergile, F.E. (2001) *Biogeography of the West Indies: Patterns and perspectives.* CRC Press.

Wunderle, J.M., Lodge, D.J., and Waide, R.B. (1992) Short term effects of Hurricane Gilbert on terrestrial bird populations on Jamaica. *The Auk,* 109, 148-166.

World Travel and Tourism Council (WTTC) (2017a) *Travel and tourism. Economic impact 2017.* Caribbean. Retrieved from: https://www.wttc.org/-/media/files/reports/economic-impact-research/regions-2017/caribbean2017.pdf on November 18, 2017.

World Travel and Tourism Council (WTTC) (2017b) *WTTC country reports.* Retrieved from https://www.wttc.org/economic-impact/country- analysis/country-reports/ on November 18, 2017.

WWF and IUCN (1997) *Centres of plant diversity: a guide and strategy for their conservation* (Vol. 3) (S. H. Davis, Ed.) Gland, Switzerland: World Wide Fund for Nature and International Union for Conservation of Nature.

Printed by Books on Demand GmbH, Norderstedt / Germany